THE SCIENCE OF LIFE

Contributions of Biology to Human Welfare

FASEB MONOGRAPHS

General Editor: KARL F. HEUMANN

Volume 1 • THE SCIENCE OF LIFE: Contributions of Biology to Human Welfare
Edited by K. D. Fisher and A. U. Nixon

A Continuation Order Plan is available for this series. A continuation order will bring delivery of each new volume immediately upon publication. Volumes are billed only upon actual shipment. For further information please contact the publisher.

THE SCIENCE OF LIFE

Contributions of Biology to Human Welfare

Edited by

K. D. FISHER
and A. U. NIXON

Federation of American Societies
for Experimental Biology
Bethesda, Maryland

FASEB, Bethesda
PLENUM PRESS, New York and London

Library of Congress Cataloging in Publication Data

Main entry under title:

The Science of life.

(FASEB monographs ; v. 1)
"The material in this book originally appeared in Federation pro-
ceedings, volume 31, no. 6, part II, November-December 1972."
Includes index.
1. Life sciences. 2. Biology. I. Fisher, Kenneth D. II. Nixon, A.U.
III: Federation of American Societies for Experimental Biology. Feder-
ation proceedings. IV. Series: Federation of American Societies for
Experimental Biology. FASEB monographs ; v. 1.
QH313.S45   574   75-5777
ISBN 0-306-34501-3

The material in this book originally appeared in *Federation Proceedings*
Volume 31, No. 6, Part II, November-December 1972

© 1972 Federation of American Societies for Experimental Biology

Plenum Press, New York is a Division of Plenum Publishing Corporation
227 West 17th Street, New York, N.Y. 10011

United Kingdom edition published by Plenum Press, London
A Division of Plenum Publishing Company, Ltd.
4a Lower John Street, London W1R 3PD, England

Printed in the United States of America

Preface

The Science of Life: Contributions of Biology to Human Welfare is the first of what we anticipate will be a series of monographs resulting from activities of the Federation of American Societies for Experimental Biology. From time to time material drawn from symposia presented at the annual meetings of the Societies, when considered suitable, will be published as separate *FASEB Monographs*. Usually, the material will have appeared in *Federation Proceedings*. Occasionally, other papers resulting from symposia, conferences, or special meetings sponsored by the Federation or one of its constituent societies will appear as a FASEB monograph. In some instances, special articles on the same topic will be drawn together under one cover.

Why should information which has already been printed and distributed as a part of the editorial content of a journal be republished as a monograph? Most of the material to be included in this effort, particularly the symposia presented at the annual meetings of the six Federated Societies, will summarize the state of the art excellently. Such information will be of considerable value to students and teachers, especially for undergraduate honors courses or in graduate studies.

In an effort to make this didactic resource readily available to students and teachers in educational institutions of all types, FASEB has arranged with Plenum Publishing Corporation to publish this new series of monographs. The marketing capability of Plenum Publishing Corporation combined with the editorial competence of the FASEB Office of Publications will make these monographs available in book shops around the world. Biologists, physicians, students, and teachers will find in these monographs an easily accessible fund of current information hitherto limited to the pages of an esoteric journal.

The first volume in the *FASEB Monographs* series tells how the results of research in the biological sciences have a bearing on our daily lives, on our health, food, environment, and the utilization of our natural resources. The text is timely because it shows, by example, how some of the problems that face us in the 1970's might be overcome through continued research on man and his environment.

The volume is a smoothly written summary of the thoughts of over a hundred scientists on this subject. From this text the reader can gain a picture of the practical accomplishments of research in biology and the need for more work.

Former Senator Lister Hill, who has had a major role in the sponsorship of medical research in the United States, provides an inspiring Foreword to this story of accomplishment.

EUGENE L. HESS
Executive Director

Contents

viii

CONTENTS

Foreword

My FATHER, Dr. L. L. Hill, early instilled in me an interest and concern for health. He named me Lister out of his admiration for the great Lord Joseph Lister, the father of antiseptic surgery, under whom my father had studied in London. I have been ever conscious of the importance of good health and the damages to mankind by disease. When the opportunity came for me to serve the people of the State of Alabama in the United States House of Representatives and later in the United States Senate, the advancement of the Nation's health became one of my major interests and concerns.

Illness is no respector of persons, of social status, or other attainments. Every segment of our population suffers when it is stricken by disease or when ill health afflicts its members. For the disadvantaged, illness and disease aggravate and often prolong, sometimes for their lifetimes, their miserable conditions of existence. A marginal financial status is pushed over the brink, poverty determining an increased incidence of disease and that, in turn, preventing the attainment of financial security. Illness in chronic or catastrophic form strikes a severe blow even at those of a more advantageous status by depleting savings, disrupting households, and deteriorating their whole life style. Only the very well-to-do can survive a lengthy illness without crippling financial results.

In the period during and after World War II good health became a national goal for all the American people. A concerted national effort was mounted to attack by research the areas of lack of knowledge which impeded medical progress and to provide hospitals where the best of modern medicine and of biological science could be brought more effectively to the care of the sick. The movement was very popular and broadly supported by the public. It was as if the country had agreed with Benjamin D'Israeli's dictum.

> The health of the people is really the foundation upon which all their happiness and all their powers as a State depend.

The progress in that period was very substantial. Poliomyelitis, tuberculosis, pneumonia, typhoid fever, diphtheria, yellow fever, and many other diseases were all but wiped out or controlled. Through research there have been advances in understanding of blindness, dental diseases, cardiovascular problems, and many other ailments; thus life has been prolonged and made more comfortable for many people.

More recently it has appeared that some of the successes of that time of strong national support for biological research, including medical research, have been overlooked. Financial support for the flourishing American research enterprise has been decreased, and training funds all but canceled. It seems that the country has turned around, that the people and their leaders have decided that biological research is not a top priority item. This decision, of course, is absolutely wrong and must not be countenanced, for research is the very foundation of improved medical care.

When the American Biology Council told me through its chairman, Dr. Edwin H. Lennette, of the proposed Task Force to document the Contributions of Biological Sciences to Human Welfare and asked me to serve as the Task Force Chairman, I was happy to do so. The meetings of these fine scientists and their discussions have been most interesting and educational. It seems to me that these dedicated men of science have produced a truly splendid and informative documentation of the contributions of research to the betterment of human welfare.

I hope that many people will review the past accomplishments of biological research as presented in this volume and will be as proud of them as I am. Let me emphasize the fact that much remains to be done and that many further contributions to human welfare are still possible if the necessary public encouragement and support are provided.

LISTER HILL
Former Senator from Alabama

Panel members

BASIC BIOMEDICINE

James W. Colbert, Jr., M.D., *Chairman,* Vice President for Academic Affairs, Medical University of South Carolina, Charleston, South Carolina

Elijah Adams, M.D., Professor and Head, Department of Biological Chemistry, University of Maryland School of Medicine, Baltimore, Maryland

A. Clifford Barger, M.D., Robert Henry Pfeiffer Professor of Physiology, Harvard University Medical School, Boston, Massachusetts

Edward W. Dempsey, Ph.D., Professor and Chairman, Department of Anatomy, College of Physicians and Surgeons, Columbia University, New York, New York

George B. Koelle, M.D., Ph.D., Professor and Chairman, Department of Pharmacology, University of Pennsylvania School of Medicine, Philadelphia, Pennsylvania

Albert R. Krall, Ph.D., Professor of Biochemistry, Medical University of South Carolina, Charleston, South Carolina

Sol Spiegelman, Ph.D., Director, Institute of Cancer Research, College of Physicians and Surgeons, Columbia University, New York, New York

Dennis W. Watson, Ph.D., Professor and Head, Department of Microbiology, University of Minnesota Medical School, Minneapolis, Minnesota

CLINICAL MEDICINE

George E. Burch, M.D., *Chairman,* Henderson Professor and Chairman, Department of Medicine, Tulane University School of Medicine, New Orleans, Louisiana

John Adriani, M.D., Anesthesiology Department, Charity Hospital of Louisiana, New Orleans, Louisiana

John C. Ballin, Ph.D., Director, Department of Drugs, American Medical Association, Chicago, Illinois

Arthur C. Beall, Jr., M.D., Professor of Surgery, Baylor College of Medicine, Houston, Texas

Paul C. Beaver, Ph.D., Vincent Professor of Tropical Diseases and Hygiene, Tulane University School of Medicine, New Orleans, Louisiana

Karl H. Beyer, Jr., M.D., Ph.D., Senior Vice President, Merck Sharp and Dohme Research Laboratories, Merck and Company, Inc., West Point, Pennsylvania

Cyril Y. Bowers, M.D., Professor of Medicine and Director of Endocrine Unit, Department of Medicine, Tulane University School of Medicine, New Orleans, Louisiana

Francis J. Braceland, M.D., Sc.D., FACP, Senior Consultant, The Institute of Living, and Editor, *American Journal of Psychiatry*, Hartford, Connecticut

John Francis Briggs, M.D., Editor, *Geriatrics*, and Clinical Professor of Medicine, University of Minnesota, Minneapolis, Minnesota

Marshall Brucer, M.D., Consultant in Nuclear Medicine, 5335 Via Celeste, Tucson, Arizona

William B. Castle, M.D., Francis Weld Peabody Faculty Professor of Medicine, Emeritus, Harvard University, and Physician, Veterans Administration, West Roxbury, Massachusetts

Russell N. DeJong, M.D., Professor and Chairman, Department of Neurology, University of Michigan Medical School, Ann Arbor, Michigan

Frank J. Dixon, M.D., Chairman, Department of Experimental Pathology, Scripps Clinic and Research Foundation, La Jolla, California

Arndt J. Duvall III, M.D., Professor, Department of Otolaryngology, University Hospital, Minneapolis, Minnesota

John P. Fox, M.D., Associate Dean, School of Public Health and Community Medicine, and Professor, Department of Epidemiology and International Health, University of Washington, Seattle, Washington

Arthur C. Guyton, M.D., Chairman, Department of Physiology and Biophysics, University of Mississippi School of Medicine, Jackson, Mississippi

Charles A. Janeway, M.D., Thomas Morgan Rotch Professor of Pediatrics, Harvard University Medical School, and Physician-in-Chief, Children's Hospital Medical Center, Boston, Massachusetts

George V. Mann, Sc.D., M.D., Nutrition Division, Vanderbilt University School of Medicine, Nashville, Tennessee

Alfred Edward Maumenee, M.D., Professor of Ophthalmology, The Johns Hopkins University School of Medicine, and Director of the Wilmer Institute, Johns Hopkins Hospital, Baltimore, Maryland

Victor McKusick, M.D., Physician, Johns Hopkins Hospital, and Professor of Medicine, Epidemiology, and Biology, Johns Hopkins University, Baltimore, Maryland

F. Gilbert McMahon, M.D., Professor of Medicine and Chief, Section of Therapeutics, Department of Medicine, Tulane University School of Medicine, New Orleans, Louisiana

John P. Merrill, M.D., Professor of Medicine, Harvard Medical School, and Director, Renal Section, Peter Bent Brigham Hospital, Boston, Massachusetts

William J. Mogabgab, M.D., Professor of Medicine, Section of Infectious Disease, Department of Medicine, Tulane University School of Medicine, New Orleans, Louisiana

Charles M. Nice, Jr., M.D., Chairman, Department of Radiology, Tulane University School of Medicine, New Orleans, Louisiana

Seymour Fiske Ochsner, M.D., Director of Radiation Therapy, Ochsner Medical Center, New Orleans, Louisiana

Arthur M. Olsen, M.D., Professor of Medicine, Mayo Foundation Graduate School, University of Minnesota, and Senior Consultant, Division of Thoracic Diseases and Internal Medicine, Mayo Clinic, Minneapolis, Minnesota

Joseph M. Quashnock, M.D., Ph.D., Colonel, USAF, MC, Commander, USAF School of Aerospace Medicine, Brooks Air Force Base, Texas

John T. Queenan, M.D., Department of Obstetrics and Gynecology, University of Louisville School of Medicine, Louisville, Kentucky

Howard A. Rusk, M.D., Professor and Chairman, Department of Rehabilitation Medicine, New York University School of Medicine, New York, New York

Walter B. Shelley, M.D., Ph.D., Professor and Chairman, Department of Dermatology, University of Pennsylvania School of Medicine, Philadelphia, Pennsylvania

Thomas E. Starzl, M.D., Professor and Chairman, Department of Surgery, University of Colorado, Medical Center, Denver, Colorado

James M. Stengle, M.D., Chief, National Blood Resource Program, National Heart and Lung Institute, National Institutes of Health, Bethesda, Maryland

Douglas D. Tofflemier, M.D., 426 17th Street, Oakland, California

John J. Walsh, M.D., Vice President for Health Affairs, and Chancellor, Tulane Medical Center, New Orleans, Louisiana

Owen H. Wangensteen, M.D., Regents' Professor of Surgery, University of Minnesota, Minneapolis, Minnesota

DENTAL SCIENCE

Joseph F. Volker, D.D.S., Ph.D., *Chairman,* President, University of Alabama in Birmingham, University Station, Birmingham, Alabama

Dwight J. Castleberry, D.M.D., M.S., Associate Professor and Chairman, Department of Prosthodontics, University of Alabama School of Denistry, Birmingham, Alabama

James M. Dunning, D.D.S., M.P.H., Professor of Ecological Dentistry, Emeritus, Harvard School of Dental Medicine, 188 Longwood Ave., Boston, Massachusetts

Sidney B. Finn, D.M.D., Professor of Dentistry (Pedodontics) and Senior Investigator, Institute of Dental Research, University of Alabama School of Dentistry, Birmingham, Alabama

Joseph A. Gibilisco, D.D.S., Chairman, Department of Dentistry, Mayo Clinic, Rochester, Minnesota

H. Perry Hitchcock, D.M.D., M.S.D., Professor and Chairman, Department of Orthodontics, University of Alabama School of Dentistry, Birmingham, Alabama

Victor Matalon, D.D.S., Associate Professor, Maxillofacial Prosthetics, The University of Texas, M.D. Anderson Hospital and Tumor Institute, Texas Medical Center, Houston, Texas

C. A. McCallum, D.M.D., M.D., Dean, School of Dentistry, University of Alabama, Birmingham, Alabama

Ralph W. Phillips, D.Sc., Assistant Dean for Research and Research Professor of Dental Materials, Indiana University–Purdue University at Indianapolis–School of Dentistry, Indianapolis, Indiana

Saul Schluger, D.D.S., Associate Dean, Graduate Dental Education, University of Washington School of Dentistry, Seattle, Washington

FOOD

Emil M. Mrak, Ph.D., *Chairman,* Chancellor Emeritus, University of California, Davis, California

R. W. Allard, Ph.D., Chairman, Department of Genetics, University of California, Davis, California

Kermit Bird, Ph.D., U.S. Department of Agriculture, Washington, D.C.

C. O. Chichester, Ph.D., Department of Food and Resource Chemistry, University of Rhode Island, Kingston, Rhode Island

M. R. Clarkson, D.V.M., American Veterinary Medical Association, Chicago, Illinois

Lawrence Claypool, Ph.D., Department of Pomology, University of California, Davis, California

Harold H. Cole, Ph.D., Professor Emeritus, Department of Animal Science, University of California, Davis, California

Victor N. Lambeth, Ph.D., Professor of Horticulture, College of Agriculture, University of Missouri, Columbia, Missouri

Dale R. Lindsay, Ph.D., Associate Director of Medical and Allied Health Education, Duke University Medical Center, and Director of Allied Health Education, Veterans Administration Hospital, Durham, North Carolina

Milton D. Miller, M.S., Department of Agronomy and Range Science, University of California, Davis, California

John W. Osebold, D.V.M., Ph.D., Professor of Immunology, School of Veterinary Medicine, University of California, Davis, California

Robert C. Pearl, Department of Food Science and Technology, University of California, Davis, California

Lawrence Rappaport, Ph.D., Department of Vegetable Crops, University of California, Davis, California

Thomas Reid, Ph.D., Professor of Animal Nutrition, Department of Animal Science, Cornell University, Ithaca, New York

Charles M. Rick, Ph.D., Department of Vegetable Crops, University of California, Davis, California

Bernard Schweigert, Ph.D., Chairman, Department of Food Science and Technology, University of California, Davis, California

Robert van den Bosch, Ph.D., Chairman, Division of Biological Control, University of California, Berkeley, California

POPULATION BIOLOGY

Frederik B. Bang, M.D., *Chairman,* Professor and Chairman, Department of Pathobiology, Johns Hopkins University, Baltimore, Maryland

John Biggers, Ph.D., Laboratory of Human Reproduction and Reproductive Biology, Harvard Medical School, Boston, Massachusetts

John B. Calhoun, Ph.D., Section on Behavioral Systems, Laboratory of Brain Evolution and Behavior, National Institute of Mental Health, Bethesda, Maryland

Paul DeBach, Ph.D., University of California, Riverside, California

William B. Jackson, Sc.D., Environmental Studies Center, Bowling Green State University, Bowling Green, Ohio

Edward F. Knipling, Agricultural Research Service, U. S. Department of Agriculture, National Agricultural Library, Beltsville, Maryland

Dwain Parrack, Department of Pathobiology, Johns Hopkins University, School of Hygiene and Public Health, Baltimore, Maryland

Lloyd E. Rozeboom, Sc.D., Department of Pathobiology, Johns Hopkins University, School of Hygiene and Public Health, Baltimore, Maryland

Brenda Sladen, Department of Pathobiology, Johns Hopkins University, School of Hygiene and Public Health, Baltimore, Maryland

ENVIRONMENTAL HAZARDS

John J. Hanlon, M.D., *Chairman,* Assistant Surgeon General, Public Health Service, Rockville, Maryland

Mary O. Amdur, Ph.D., Associate Professor of Toxicology, School of Public Health, Harvard University, Boston, Massachusetts

D. L. Alverson, Ph.D., National Marine Fisheries Service, East Seattle, Washington

Philip A. Butler, Ph.D., Gulf Breeze Laboratory, Environmental Protection Agency, Gulf Breeze, Florida

A. C. Jones, Ph.D., Southeast Fisheries Center, National Marine Fisheries Service, Miami, Florida

Charles E. Lane, Ph.D., School of Marine and Atmospheric Sciences, University of Miami, Miami, Florida

John J. Magnuson, Associate Professor, Laboratory of Limnology, Department of Zoology, University of Wisconsin, Madison, Wisconsin

J. L. McHugh, Ph.D., Marine Sciences Research Center, State University of New York, Stony Brook, New York

Ross F. Nigrelli, Ph.D., Osborn Laboratories of Marine Sciences, New York Aquarium, Brooklyn, New York

Brian J. Rothschild, Ph.D., Director, Southwest Fisheries Center, National Marine Fisheries Service, La Jolla, California

J. L. Runnels, Ph.D., Acting Chairman, Division of Fisheries and Applied Estuarine Ecology, RSMAS, University of Miami, Miami, Florida

John H. Ryther, Ph.D., Woods Hole Oceanographic Institution, Woods Hole, Massachusetts

Carl Sindermann, Ph.D., Middle Atlantic Coastal Fisheries Center, National Marine Fisheries Service, Highlands, New Jersey

Maynard A. Steinberg, Ph.D., National Marine Fisheries Services, Pacific Fishery Products Technology Center, Seattle, Washington

NATURAL RESOURCES

Frederick Sargent II, M.D., *Chairman,* School of Public Health, University of Texas, Houston, Texas

Thadis W. Box, Ph.D., Dean, College of Natural Resources, Utah State University, Logan, Utah

Vernon Bryson, Ph.D., Institute of Microbiology, Rutgers University, New Brunswick, New Jersey

C. H. Driver, Ph.D., College of Forest Resources, University of Washington, Seattle, Washington

E. J. Dyksterhuis, Ph.D., School of Natural Biosciences, Department of Range Science, Texas A & M University, College Station, Texas

R. I. Gara, Ph.D., College of Forest Resources, University of Washington, Seattle, Washington

S. P. Gessel, Ph.D., Associate Dean, College of Forest Resources, University of Washington, Seattle, Washington

P. L. Johnson, Ph.D., Division Director, Environmental Systems and Resources, National Science Foundation, Washington, D.C.

THE SCIENCE OF LIFE
Contributions of Biology to Human Welfare

Chapter I

Introduction

EDWIN H. LENNETTE, M.D., Ph.D.
AND J. F. A. McMANUS, M.D.

> History is a fragment of biology: the life of man is a portion of the vicissitudes of organisms on land and sea ... Therefore the laws of biology are the fundamental lessons of history. We are subject to the processes and trials of evolution, to the struggle for existence and the survival of the fittest to survive. If some of us seem to escape the strife or trials it is because our group protects us; but that group itself must meet the test of survival.
>
> Will and Ariel Durant
> *The Lessons of History*
> New York:
> Simon & Schuster, 1968

THIS BOOK DEALS with the health and survival of man. It leaves many things unsaid. It is not comprehensive—no single volume can possibly be. The intent is to present the major contributions over the past three decades of biomedical research to the welfare of man.

In the late 1960's, the national scientific endeavor of the United States, until then enthusiastically supported and unquestionably accepted, became the target of criticism from a number of sources. Science and technology were accused of being extravagantly, even wastefully, financed and of causing damages which overbalanced their benefits.

The dominant influence of science and technology is perhaps the outstanding fact of the world of today. Its influence extends over every aspect of human life whether

1

it concerns food production, public health, heavy industry, communications, commerce, entertainment, the operation of our own homes, politics, or religious belief. Both the credit and the credibility bestowed on modern science rest, however, not on an appreciation of concepts but on technological accretion and complication. The technical applications stemming from research are often confused with science itself. That a host of advances in biology had greatly reduced many diseases, some to the point of eradication, and that the existence at a marginal level of large segments of the population of the country was no longer necessary, were almost forgotten. It seems to biologists that there is a need, and indeed a specific demand, for a documentation of the contributions of biology to human welfare.

The American Biology Council, formed by the Federation of American Societies for Experimental Biology and the American Institute of Biological Sciences, two organizations whose membership comprises the majority of American biologists, established a Task Force in 1969 for this purpose. Under the joint chairmanship of Senator Lister Hill and Dr. Edwin H. Lennette, then president of the American Biology Council, it was planned that this document would consider some of the most recent findings of biology and indicate how the welfare of man has been improved. The report would be in a form readable and comprehensible to the nonscientist.

Eminent scientists served as Panel Chairmen and together they comprised the Task Force. They were responsible for the selection and enlistment of the Panel members, for providing adequate coverage of subjects within their areas, and for preparing a synthesis of the contributions of the Panel members.

Each author is a leader in his subject, working at the laboratory bench or in the hospital wards or in the field, developing new knowledge and applying it to the advancement of human welfare. This introductory chapter outlines the material covered and indicates a few specially significant accomplishments.

The body of the book, Chapters II to IX inclusive, documents dramatic beginnings of a recent and continuing revolution in biology. Numerous discoveries combine to provide a foundation on which grows an ever-increasing understanding of man's needs and his relationship to environmental factors and influences. The workers have been legion, the studies have been extraordinary in number and variety, and an entirely new view of man and nature is emerging. Only a sample of the most significant findings regarding man, his health, his environment, his food, the hazards, and the natural resources on which life depends are presented.

Biology, the study of living things, is concerned with the origin, development, structure, and functions of plants and animals. Despite a long evolutionary history, only since the time of Aristotle, who lived six centuries before Christ, have biological observations been recorded. During the 2,500 years from these first writings until today, studies in biology have multiplied greatly in number, intricacy, and complexity. The direction has proceeded from studies based on the whole animal to studies concerned with the individual molecules that make up the tissue and organs of the animal. Generations of biologists have examined, described, and classified organisms and parts of organisms. Despite their efforts only a small fraction of all the different species of living organisms have been classified. Paleontological evidence indicates that countless other species of plants and animals have evolved and perished during the past millions of years. Many existing species presently unclassified may be potentially useful or possibly hazardous to humans.

As knowledge about the nature of living things, that is the flora and fauna living on this planet, improved, the interests of biologists shifted from descriptive to functional matters. The transition in the direction of biological and medical research resulted from an increased understanding of basic principles. While the transition is recognized by biologists it is important that it be recognized by everyone. Many, perhaps most, people are unaware that the structure of society and the relationship of human beings to their environment is to a large measure determined by principles of biology discovered within the past 50 years. Because of these discoveries civilization and the status of mankind in the world of the future will be vastly different from what it is today.

In this book you will find what biologists have learned about man. How he is constructed. How he can remain healthy. How he can conquer sickness. How he resembles other creatures. The story also discusses man's relationship with his environment, how he can use the natural resources of the land and the sea and the havoc he can cause by misusing these resources. We attempt to chronicle recent advances in biology in hope that the effort will be useful to those who face the problems and opportunities of the future. Human experience suggests that a view of the past provides an insight into the future.

Science in general, and biology in particular, have flourished in the Western countries in the past quarter century. This has resulted in large measure from an intelligent use of public and private money available for supporting a wide diversity of scientific and technological activities. While involved in this endeavor, the scientist has made little effort to improve public understanding of the nature of discovery, or to establish connections between what science discovers and what technology uses. Social needs and aspirations sometimes appear to have been neglected. A few articulate and persuasive voices have challenged the precept of science as a universal good. Some recent advances in biology and medicine have been identified as being ethically bivalent; that is, potentially useful and potentially dangerous. A reading of social history discloses that such criticism has been directed to nearly every technical discovery ever recorded.

Medical sciences are devoted to man's health and welfare. The biological disciplines intimately concerned with man and his illnesses form the foundation on which modern medicine operates. Before effective treatment is possible, we must know how the body works and how its function is affected by disease or by the drugs used in treatment. Such studies are generally designated as Basic Biomedicine (Chapter II).

The clinical sciences also form a particularly important and interesting area of biology. In Chapter III, Clinical Medicine, the conquest of tuberculosis, syphilis, and many other plagues of human history, resulting from the discovery and application of antibiosis, provides an engaging tale of the elimination of scourges of the past and our successes in understanding and reducing those which remain.

Until the year 1650 the average life expectancy for humans at the time of birth was never more than 25 years. For thousands of years the lament of an ancient Chinese poet

Like fallen leaves, they tumbled to the Nether Springs
All vanished ere their middle years were passed.

described man's future. At the beginning of this century the average American could look forward to 49 years of life, and many diseases about which we hear little today were matters of great concern. Diseases of bacterial origin decimated the ranks of the living. Effective treatment for lobar pneumonia, tuberculosis, peritonitis, meningitis, pancarditis, and many other diseases was nonexistent. Hospital beds were filled with patients ill with infectious diseases of one kind or another.

In 1935 a new drug called "Prontosil" was announced to the world. Within a year the active principle of the drug was found to be a rather simple substance called diaminoazobenzene sulfonamide, well known to organic chemists. Biologists had been given a guidepost and innumerable "sulfas" such as sulfadiazine and sulfathiazole came out of the laboratories. Deaths from many fatal diseases became uncommon and man's average life-span was prolonged.

The development of penicillin initiated the era of wide spectrum antibiotics that gave powerful tools to treat effectively many diseases, shortening their course, restoring the patient to health, and enabling him to become economically productive in a comparatively short time. Syphilis became amenable to quick cure, and congenital syphilis and syphilitic heart disease became rare. The incidence of rheumatic fever, a notorious contributor to heart disease, was tremendously decreased and the incidence of other streptococcal diseases showed a marked decline.

Penicillin, a by-product of World War II, continues to save more lives each year than were lost during that entire conflict. Over the past 30 years antibiotics have become big business, with a total production of nearly 17 million pounds in the United States alone in 1970. More than 145 million dollars was spent on antibiotics in the United States during that year.

Almost immediately following the availability of penicillin the chance observations of a soil microbiologist led to the discovery and development of streptomycin. This antibiotic has been used effectively in the treatment of tuberculosis. During the period 1954–1969 savings from decreased hospitalization costs for tuberculosis patients have been estimated at 3.77 billion dollars, and the increased productivity of former tuberculosis patients has been estimated to amount to 1.2 billion dollars. The discovery in 1944 of a particular soil organism led to an estimated economic benefit of 5 billion dollars between 1954 and 1969. Nor does this include the intangible value of the human lives themselves.

The information gained in eliminating many bacterial diseases is now being applied to diseases caused by viruses. Within the past 25 years hundreds of new viruses have been discovered many of which have been implicated in human, animal, and plant diseases. The development of new vaccines and the improvement of older ones have occurred during this same period. The vaccines prevent diseases rather than treating diseases that have occurred. Highly effective vaccines protect against poliomyelitis, measles, and rubella. Knowledge about the influenza viruses has led to a development of some vaccines and a mechanism of surveillance that detects the appearance of new strains of virus allowing appropriate vaccines to be prepared in advance of an epidemic.

The only association that the average person has with these developments is the injection into his arm of a

small quantity of material, usually when he is young or
when he is exposed to the disease. While advances are
often dramatically and graphically described in news-
papers the descriptions often overlook or ignore the
long, arduous, patient efforts of large numbers of researchers that have contributed to the development.

Poliomyelitis has been virtually eliminated within
the past 15 years. An important finding that led to the
production of poliomyelitis vaccines was the observation
that poliomyelitis viruses could be grown in the laboratory in cells from monkey tissue. While the research that
led to this discovery was actually directed to the role of
viruses in cancer, the finding was applied by others to
poliomyelitis. It is not possible to estimate the benefits,
both social and economic, that have flowed from the
successful application of basic discoveries to the elimination of poliomyelitis as an annual summer threat to the
lives and happiness of thousands of families. Various
estimates of costs and savings have been published, all
in the billions of dollars. Apprehension and dread is no
longer present in the hearts of countless mothers who
feared the paralysis and death of their children as polio
epidemics scourged the country year after year.

The virtual elimination by the end of the first half
of this century of most bacterial infections is one of man's
crowning achievements. But the conquest of disease or
any other condition adversely affecting the welfare of
man, while it may be gratifying, does not put an end to
the problem. Constant vigilance must be maintained
to meet the possible appearance of unforeseen complications and consequences. Emptying an ecological niche
occupied by one disease often provides an opportunity
for other disease agents to emerge or for unrecognized
pathogens to be identified.

A disease process called subacute sclerosing panen-
cephalitis, also known as Dawson's encephalitis, arises
from activation of a latent virus, and afflicts children and
young adults primarily. Evidence indicates that some
individuals who acquire measles in childhood and re-
cover continue to harbor the virus in a latent form.
Years later, by a mechanism still unknown, the viruses

become reactivated and initiate destruction of cells in the brain. The etiology, which only a few years ago was unknown, is now understood and therapy can be prescribed on a somewhat more rational basis than was possible without this knowledge.

A central nervous system disease known as Creutzfeldt-Jakob disease, a presenile dementia, has been shown recently to be a chronic degenerative process caused by a virus. The finding is so new that the virus has yet to be characterized as to its nature and properties. Such discoveries encourage hope that the etiology of other diseases, for example, multiple sclerosis and amyotrophic lateral sclerosis, may yield etiological clues to the prying and probing of the many bioscientists engaged in such studies.

Organ transplantation, especially that of the kidney and heart as described elsewhere in this book, represents a major achievement of surgery. However, as a prominent biologist recently commented, we have "such examples of 'half-way technology' as the artificial kidney and renal transplantation for the treatment of kidney failure, while the underlying cause of the disability—most frequently chronic glomerulonephritis—remains essentially unsolved. It is predictable that when a clearer understanding has been gained of the mechanism by which immune precipitates are deposited within the walls of the renal glomeruli, when the precise nature of the antigens involved has been identified, and when the disease can be approached by measures designed to prevent or reverse these phenomena, there no longer will be a need for the substitution or replacement of human kidneys."

Economic and technical burdens associated with such interim measures as kidney dialysis, for example, cost many thousands of dollars per patient per year, impossible to finance for thousands of individuals with kidney failure. Heart transplantation requires large teams of highly skilled surgeons, physicians, and scientists. If several hundred thousand cardiac transplants were done annually, a reasonable estimate of the potential cases, the costs alone would be overwhelmingly large.

The answer obviously is not palliative treatment but
discovery of mechanisms for either prevention or cure
of the primary disease process.

The molecular approach to biology exposed an entire
new spectrum of diseases called inborn errors of metabolism. A defect in a gene may give rise to a defective
or nonfunctional enzyme. Depending on the degree of
defectiveness and the role of the enzyme, the result may
be either trivial or important to the health and well-being
of the individual carrying the defective gene. It is possible
that almost all individuals suffer from one or more
inborn errors of metabolism most of which are incon-
sequential to normal living. At least 130 such abnormalities have been identified so far.

Dental science, discussed in Chapter IV, applies the
results of basic biological research to dental diseases that
afflict approximately 80% of the adult population. When
individuals survive the diseases of childhood a greater
fraction of individuals attain adulthood. They require
space, oxygen, food, and continual medical attention
also. Dental science, for example, has established the
factors that cause and accelerate the development of
dental caries and periodontal disease. Techniques have
been developed that will reduce the incidence of tooth
decay. The sheer number of individuals with dental
problems requires massive attention to and correction
of dental diseases.

Science is a by-product of agricultural surplus. It is a
contribution along with such things as poetry, literature,
art, and music of that proportion of the population not
required for food production. The United States now
produces food surpluses using no more than 10% of its
population. In this country we can now spare a large
fraction of the population for the production of bathtubs,
refrigerators, television, golf clubs, snowmobiles, and
other luxuries and conveniences of civilized life. Not
only do we enjoy the great surpluses of agricultural
products but also the problems associated with such
surpluses.

The methods used by the farmer in 1900 did not differ
greatly from those of previous centuries except for the

introduction of the horse-drawn steel plow, the reaper, and a few other simple machine tools. The farmer of 1900 would not recognize the practice of farming as carried out in 1972. Indeed most people are unaware of the revolution in farming methods that has occurred during the past 20 years. The introduction of specialized motor-driven tools, of scientifically developed strains of crops and livestock, of new irrigation techniques, of chemicals to control weeds, of pesticides, of new fertilizers and fertilization techniques, and chemicals for the control of growth and disease in poultry and cattle; all of these changes occurring as a by-product of science have allowed great expansion in productivity in the face of a sharply declining farm population.

The American farmer today can produce enough food to feed himself and 42 other people; of these 42 persons, five live outside the United States indicating the extent to which efficient agricultural practices in this country can feed the rest of the world. Application of scientific agricultural techniques designed and worked out in "developed" countries, and particularly in the United States, have led to phenomenal increases in the production of food grains in underdeveloped countries with large populations. Even in the face of this greatly increased food production, the present "Green Revolution" can buy perhaps only 20 more years for a world faced with overpopulation. Increasing populations have challenged the biologist to increase the food supply to meet an ever-growing demand.

All of us are beneficiaries of these research efforts. All of us benefit indirectly by the low incidence of those diseases which have been brought under control and directly by the improved preventive and therapeutic medical care that is available and that we seek actively when required.

An important aspect of the reduction in infectious diseases has been the emergence of an increasingly large population of human beings. More and more adults are surviving to extended ages. Where death of children was commonplace only a few years ago, now it is the exception, and survival to old age is taken for granted. The

change has led to a reorientation of our approach to
medical care. The problems associated with advancing
age and the delivery of adequate health care to older
people have become of paramount importance. At the
same time the survival of a larger fraction of individuals
to puberty increases the rate of population growth.

The biology of populations considered in Chapter VI
is a relatively new area of study. Workers in this field
gather and collate information requisite to an improved
understanding of the behavior of the human population.
Chapter VI surveys human populations but also those
insect and animal populations that can be examined
readily when environmental conditions are controlled
experimentally. Population biology is a rapidly developing field which promises much for the future, perhaps a
whole new way of looking at living things: plants, animals, and man. The studies already have contributed
to an understanding of the influence of population density
on behavior.

The modern environment, as described in Chapter
VII, Environmental Hazards, has become dangerous.
Air pollution is a result of neotechnical culture. Along
with industrial development came the firing and the
open venting of fossil fuels such as oil, coal and gasoline
in humid and semihumid climates releasing millions of
tons of corrosive substances into the atmosphere. Those
corrosive materials that return to the ground through
rain and fallout, downwind from the sources of pollution,
often inflict devastating damage to animal and plant life
and to stone and other man-made structures in urban
areas. Emphysema, a fairly rare occupational disease a
few decades ago, has now become a commonplace
product of air pollution. The exhaust from automobiles
is one of the worst offenders.

The environment has been changed particularly
rapidly in recent years. The indiscriminate use of pesticides and the unrestricted dumping of wastes into rivers,
lakes and oceans have endangered animals, birds, and
fish.

While pesticides have been of great assistance in the
effort to produce more food for more people, experience

has emphasized the deleterious effects of some of these useful agents. Biological control of pests is being developed, as well as more selective and less persistent chemical pesticides.

Methods for efficient utilization of the rangelands which will maintain the herds of domestic animals required for man's sustenance require the preservation of whole ecosystems. If man is to survive on this earth, it will be because of basic research of the variety and depth described in Chapter VII.

The Marine Sciences discussed in Chapter VIII describe the inhabitants of the oceans, estuaries, lakes, rivers, and streams. Half of the daily requirements of food protein for the human population could be secured from the seas. It is essential, therefore, that research into marine food production be pursued. The annual marine fishery harvest of about 60 million metric tons, which approximates in weight the world production of meat and poultry, emphasizes the importance of the sea as a source of food. Development of methods for the location of fish, based on knowledge of fish behavior, has increased the fish harvest but has shown the need also for management of marine resources. Overharvesting can lead to the reduction of useful resources as much as can pollution.

The role of the environment on materials and resources useful to man is described in Chapter IX. Many of our natural resources are renewable, fortunately. The replenishment, or replacement, of the resources on which survival depends requires that they be used with care and managed wisely, and this requires an intimate understanding of ecological relationships. Some of our most useful and important natural resources are seriously threatened; their preservation or restoration will depend on the intelligent application of basic principles, some of which are known, and others yet to be discovered.

A factor of great significance to human welfare was the recognition that biological structures are organized on a molecular basis. The molecular approach to biology that developed slowly during the first part of the 20th century has provided a unifying concept that has had

profound influence on all aspects of biology and medicine. The fruitfulness of the molecular concept in stimulating biomedical research will be apparent throughout
the book.

As the concept grew and flourished during the 3rd,
4th, and 5th decades of the century and as new and
incisive research instruments and techniques became
available, innumerable experiments demonstrated the
unity of structure and processes throughout all living
things. Identical or remarkably similar molecules and
mechanisms were found in the cells of bacteria, yeast,
molds, and men. The same code was used by plants and
animals to store and to transmit information. Experimental results obtained cheaply and speedily employing
bacteria or small animals as models could be used with
confidence to predict biochemical events occurring in
the human body. Although holistic attitudes dominated
biology at the beginning of the century, by midcentury
the reductionist had taken over. The reductionist approach accomplished what holism could not, and established the unity and probable kinship of all creatures.

As the biologist has examined an ever larger number
of different organisms and probed ever further into the
innermost secrets of life itself, with increasingly incisive
tools and techniques, his confidence in eventually understanding the processes of life grows stronger. Associated
with an understanding of the processes is the ability to
manipulate them to improve the welfare of all mankind.

Chapter II

Basic biomedicine

JAMES W. COLBERT, JR., M.D.

In the early days of Western science some investigators, who were called systematists, classified organisms into taxonomic groupings. Anatomists studied the structure of living tissue and physiologists were concerned with the function of the structures. The areas of anatomy and physiology were never completely isolated from each other. Chemists and physicists, who studied nonliving matter, became less separated from biologists and medical scientists as time went on. Prior to the invention of the microscope, anatomists could study only those structures that could be seen with the naked eye or perhaps with the aid of a simple lens. With the invention of the light microscope, they delved into things beyond ordinary vision. Not only parts of organisms but entire organisms too small to be seen by eye alone could now be investigated. In very recent times, with the invention of the electron microscope and X-ray techniques, scientists have penetrated to the very molecules and atoms of which living matter is made.

The physiologists, meanwhile, investigated the way in which these smaller and smaller units functioned. By now some chemists had become biochemists, concerned with the chemical makeup and chemical changes in living things. Physicists also became interested in biological problems and the field of biophysics came into being. Anatomy and physiology spawned subdisciplines, such as renal physiology, physiology of reproduction, neurophysiology, and so forth. Today the various disciplines and research areas are interrelated and overlapped

15

in the most intricate and intimate ways. During the past 30 years along with specialization have come outstanding and sometimes astounding achievements of greatest importance to medicine and human health.

Just as there are many reasons why an automobile may perform at less than optimum efficiency, e.g., a flat tire, a defective spark plug, or a dirty carburetor, there are many reasons why a human body may be out of order and a person feels sick. Basic biomedicine is concerned with understanding the normal processes that function properly in good health and in defining malfunctioning mechanisms that result in ill health.

SPEED OF APPLICATION

Until quite recently the most important factor limiting research progress has been inadequate experimental techniques. Today, on the other hand, progress is more dependent on new ideas and new theoretical concepts. An idea or a technique requires much less time now than formerly to become applicable in medicine and medical practice. Although William Harvey, the 17th century scholar who has been called the Father of Quantitative Physiology, lived to the advanced age of 79 years, he did not live long enough to see the impact on clinical medicine of his revolutionary studies on the circulation of the blood. The situation was different with the work of physiologist Walter B. Cannon. Only a year after the discovery of X rays, Cannon began using the new X-ray tube to study the passage of material through the gastro-intestinal tract. In December 1896 he observed food as it went down the esophagus in a variety of animals. In 1897 he studied the process of swallowing in man. In 1905, a medical scientist used the X-ray technique to complete a thorough study of the human alimentary canal. In a short period of 10 years the use of X rays had spread from the physics laboratory to the physiology laboratory and then to clinical medicine. Similar examples can be cited in most areas of medicine. Within a year of the laboratory synthesis of hormones called prostaglandins, clinical trials demonstrated their use in terminating pregnancy in humans.

The investigation of reproductive physiology has contributed not only to the treatment of reproductive disorders, but also to population control, a major contemporary problem throughout the world. Research in physiology and biochemistry led to a clearer understanding of the female reproductive cycle and to the development of the "pill." The "pill," discussed also in another chapter, resulted from a better understanding of the way in which hormones regulate the menstrual cycle. Although current varieties of the "pill" are not completely satisfactory, new means of biological contraception are being developed that promise to displace those now available. A short-term pill that will induce regular menstruation, regardless of sexual activity, seems certain within the foreseeable future, an event made possible by the recently discovered prostaglandins.

The prostaglandins induce abortion in both animals and man, when given intravenously early in pregnancy. When clinically warranted, it may soon be possible to take prostaglandins orally or perhaps use a vaginal pessary to intercept development of the fetus. Postcoital contraception may provide advantages over other methods now in use since tension and disturbing psychological factors would be eliminated. The undesirable side effects of the currently available "pill" are not anticipated. Clinical trials testing the efficacy of this method and looking for adverse side effects are already under way.[1]

REGULATORY MECHANISMS

The existence of and the necessity for subtle highly organized mechanisms for coordinating and controlling the activities of living organisms have been recognized since antiquity. The most complex of these appear to coordinate processes of the brain and nervous system. More primitive, perhaps, are processes which regulate

---

[1] Since the prostaglandins are not contraceptives but rather abortifacients, added serious ethical considerations are involved.

and maintain a consistent internal environment. In the case of vertebrate animals blood plasma bathes the cells and tissues. The action of the lungs and kidneys maintains the concentration of salts and a mild alkalization of blood plasma with remarkable consistency. Insulin, one of the hormones, acts to regulate and control the level of sugar in the blood. Other mechanisms for controlling metabolic processes, the most recently discovered and a most active area of biochemical research, will be discussed later.

An interesting and important discovery is the mechanism whereby the central nervous system influences the reproductive cycle and most of the events associated with reproduction and lactation. The breeding season for wild animals is known to be geared to the physical environment and more specifically to the changing, usually increasing, length of daylight. Vision, sound, sexual odors, and the mating act also influence reproductive functions. How do nerve impulses in the brain, which are instigated by these sights, sounds, and smells, get to the pituitary gland at the base of the brain, or cause other stimulating messengers to get there? Chemical messengers, releasing factors, are transmitted from the brain to the pituitary by a special set of blood vessels. The releasing factors enter the gland and stimulate the secretion of hormones that in turn govern functions of the ovaries and the testes. The chemical structure of one of these messengers has been determined and that of others is nearing realization. When these substances become available in the clinic, physicians will have new and positive means of controlling fertility and overcoming sterility in the human. Planned conception, maintenance of pregnancy, and uneventful labor are important prerequisites for the well-being of mother, fetus, and child.

The regulatory process that governs the functioning of the uterus is understood in general terms. The uterus is an organ that holds the fate of more than 3.5 million babies born within the United States each year. The uterus remains relatively quiescent during the 280 days of fetal development and suddenly at the end of this gestation period, quickly expells its contents. The complex regulatory system controlling this behavior includes certain sex hormones. Progesterone, for example, helps keep the uterus quiet until just before expulsion occurs. Other female sex hormones called estrogens develop the contractible elements in the uterus. The contractions, although painful, are needed for successful expulsion of the baby. Another hormone, oxytocin, released from the brain triggers the uterine contractions known as labor pains.

Information acquired after many years of study about the complex and changing food requirements of developing egg and sperm cells has enabled fertilization and culture of mammalian egg cells outside the body. Fertilization of the egg has been accomplished outside the body of a mouse and the egg has been transferred to a receptive host mouse. Instruments have been developed that can be inserted into the abdomen of the human female to collect ripe human eggs without surgery. Eggs from humans have been fertilized outside the body and made to develop beyond the 16-cell stage. Transfer to a receptive host has not yet been attempted in humans, but the achievement in mice represents a major advance toward overcoming sterility in women caused by blocked egg passages. The application of these and other techniques to animal husbandry can lead to important improvements in agricultural productivity.

THE LUNGS

The modern era of pulmonary physiology started from practical goals in aviation medicine. A great deal of additional basic knowledge of the respiratory system was required to attain the goals. The basic studies in pulmonary physiology have become cornerstones of current medical practice in chest clinics and in catheterization laboratories. The studies have contributed also to anesthesiology, industrial medicine, diving medicine, hyperbaric medicine, and to the diagnosis and treatment of acute and chronic respiratory infections. Despite the substantial gains in knowledge more than 100 persons continue to die every day from chronic respiratory diseases, and some 50 infants die daily from acute respiratory disorders. Much additional information about respiratory function is needed.

The lung is a major airway, a tube that, after about 23 successive divisions, terminates in some 300 million discrete microscopic sacs called alveoli. These alveoli provide about 600 square feet of surface in which oxygen from inhaled air is exchanged for carbon dioxide in blood carried in capillaries, the smallest of the blood vessels. Across this membranous surface inhaled oxygen moves from the air in these tiny sacs into the blood to be carried to the tissues of the body by the hemoglobin in the red blood cells. Carbon dioxide from the tissues moves from the blood into the air within the sacs, to be exhaled from the body in the breathing process.

The large alveolar surface is accessible also, unfortunately, to hazardous agents carried by the air, such as noxious gases and particles, as well as to beneficial and anesthetic gases administered in medical treatment. The great expanse of alveolar lung surface is the major interface between the atmospheric environment and the blood. Changes in this delicate arrangement can lead to an insufficient supply of oxygen for the blood and tissues and to an inadequate removal of carbon dioxide. Undesirable changes can arise from mechanical interference with the pulmonary airway as in asthma and emphysema, or from improper distribution of blood to the capillaries, or both. Malfunctions such as silicosis result from an accumulation of foreign particles. Other common ailments are a tissue derangement, fibrosis; atelectasis, a respiratory distress disease of the newborn resulting from a collapse of the alveolar sacs; and acute infectious diseases such as bronchitis, pneumonia, and tuberculosis.

It has become possible to describe mathematically and precisely the behavior and distribution of gases deep in the millions of alveolar sacs; and this has led to the realization that many of the conditions of inadequate blood oxygenation can be attributed to an uneven

distribution of air in the alveoli and of blood in the alveolar capillaries. These concepts led to success in assessing,
by measurement of the oxygen difference between the
air in the alveoli and that in the arterial blood, the
efficiency with which the lungs transmit oxygen to the
blood, now a major tool in medical diagnosis.

An interesting development in understanding problems of gas exchange between alveolar air and blood
was the finding that when in the upright position the
upper part of the human lung receives less air and less
blood than when prone and receives them in unequal
proportions.

A new era in pulmonary mechanics came with an
understanding of the behavior of the elastic structures
of the lungs and chest wall, and of the role of surfactant,
a substance that coats the alveolar walls. The development of modern devices for artificial respiration, for
delivering anesthetic gases, and for underwater breathing, such as scuba diving, are products of basic research.
New concepts and tools have led to improved methods
for the quantitative assessment of the mechanical factors
affecting the flow of gas in and out of the lungs. It has
been found that a substance called surfactant is secreted
by the lung tissues and must be present to maintain the
mechanical stability of the millions of tiny alveolar sacs.
Without surfactant the alveoli collapse. The discovery
accounts for a respiratory syndrome of newborn children
where surfactant is either missing or inadequate.

The discovery that high concentrations of oxygen are
hazardous, and cause blindness in premature infants
kept in incubators, has led to the proper use of oxygen in
hyperbaric chambers used for treatment of carbon
monoxide poisoning and gas gangrene. Recent evidence
suggests that oxygen may delay or temporarily reverse
cerebral deterioration in the aged, a finding important
to understanding senility. The problems of aging and of
preventing premature physiological deterioration have
received only meager attention.

THE HORMONES, INSULIN, AND DIABETES

Some cells of the body such as those in reproductive tissue do not function in the absence of hormones and waste away. The reproductive system provides a particularly favorable situation for the study of complex hormone action. The sex hormones can be labeled with radioactive carbon or hydrogen atoms, making it possible to identify whether the site of action is on the cell membrane, on constituents within the cytoplasm or in the nucleus. Incisive and efficient techniques have been developed for isolating components from within the cell. The effect of hormones on biochemical properties can be studied better on the isolated material. The information can be of incalculable benefit to the physician when prescribing hormones. A knowledge of the structure and function of hormones followed by analysis and synthesis has permitted more careful evaluation of their functions and important improvements in clinical medicine. The findings contribute to a better understanding of the aging process and of the so-called degenerative diseases, and to improved means of dealing with these matters.

Biologists are beginning to comprehend how extremely minute amounts of a hormone can trigger profound physiological changes. An important and recent advance has been the discovery of a cellular intermediate called cyclic AMP that evidently, in a wide variety of cells, serves as a regulator of cellular processes. Many hormones seem to function by stimulating cells to make more of this intracellular mediator.

Although in 1940 insulin had been in use for almost 20 years, little was known of the mechanism of its action. As the use of insulin greatly decreased the number of deaths from diabetic acidosis, an increasing percentage of deaths among diabetics resulted from vascular complications, notably heart attacks and kidney failure. Diabetics whose lives were prolonged by insulin were suffering vascular damage to the retina of the eye and nerve damage. The intimate relationship of diabetes to carbohydrate metabolism was recognized but its relationship to fat metabolism was not understood.

A major advance occurred when isotopes became
available as metabolic tracers. The use of radioactive
or stable heavy isotopic tracers whose distribution can be
followed by physical methods provides a remarkably
effective method for following complex metabolic events.
Although a product of discovery and research in physics,
isotopes were applied to biomedical problems as early as
1934. Isotopic tracer methodology developed rapidly
after World War II when isotopes became available in
quantity and variety at a reasonable price. The methodology has had an enormously useful influence on almost
all facets of biology and experimental medicine.

Metabolic tracers were used to show that body fat,
instead of being an inert food reservoir, is a very active
metabolic tissue, and the source of acid bodies carried
in the blood of the diabetic. Diabetes is now known to
be a disease associated with the metabolism of both sugar
and fat (lipid). Research has revealed in part the mode
of transportation of lipid in the blood. Methods have been
developed for determining the concentration of lipids in
blood plasma and the search continues for methods to
prevent cardiovascular complications. The information
about the physiological and biochemical changes that
occur in diabetes has improved therapy in diabetic
acidosis with measurements of glucose, acid bodies, and
minerals helping to assess the status of the patient and
his needs for insulin, carbohydrate, fluids, and minerals.
An interplay in diabetes between the pituitary gland
and the pancreas has been discovered also, and surgical
removal of the pituitary has been used to reduce vascular
degeneration in the retina. The second most frequent
cause of adult blindness in the United States is vascular
degeneration in the retina of diabetics.

Further advances in understanding diabetic disease
will result from: 1) determination of the site of action of
insulin and more precise knowledge of its mechanism of
action, which should lead to new agents for treatment
2) identification of ways to detect those predisposed to
diabetes, which may open the way to recognizing the
underlying malfunction; 3) increased knowledge about
vascular effects, which may help reduce or prevent

complications in the overt diabetic; *4)* improved methods for assaying insulin, leading to better clinical control of the disease.

HORMONES AND ULCERS

Although hormones were discovered first in the alimentary canal, only recently, with the isolation in pure form and synthesis of the gastrointestinal hormones, have investigators been able to unravel the complexities of their actions. Each hormone has multiple actions that interact to control a sequence of secretions in the digestive tract. The role of nerves in controlling the secretion through release of hormones has become clearer also. Hormones, nerves, and gastrointestinal secretions are involved in complex feedback systems that control gastrointestinal secretory and mechanical activities. The term "feedback system" describes regulatory systems exemplified by thermostats and other control devices. An improved understanding of the regulating process may lead to better methods of treating peptic ulcers. Ulcers occur when the digestive tract fails to protect itself against its own digestive juices.

There are two facets to the ulcer problem: *1)* attack against the wall of the digestive system by hydrochloric acid and pepsin, and *2)* defense of the wall against these by its inner lining, the mucosa. An excessive secretion of the hormone called gastrin stimulates an excessive secretion of acid. An abnormally high con- centration of acid overwhelms the acid-neutralizing capacity of the mucosa and the digestive wall becomes eroded with ulcers. The wall of the small intestine is a frequent site for ulcers. In addition to ulceration, high acidity in the small intestine interferes with the digestion and absorption into the blood stream of fat, leading to undernutrition and diarrhea. An improved understand- ing of how gastrin stimulates acid secretion may lead to additional methods for control of ulceration.

A conspicuous achievement in physiology has been the elucidation of the cause of pernicious anemia, and a means of treating this disease. Investigation revealed that patients with the disease fail to produce a particular

protein essential for the absorption into the blood of
vitamin B_{12}. With an inadequate absorption of vitamin
B_{12}, the production of red blood cells decreases. The
result is a chronic progressive anemia that is known as
pernicious anemia.

Many problems of gastrointestinal physiology remain
unsolved. Inadequate absorption of metabolites from the
digestive tract results in malnutrition that still plagues
large segments of the population. We are beginning to
understand some aspects of the malabsorption, especially
that resulting from genetic disorders. As additional
information becomes available an understanding of
gastrointestinal function in health and disease will improve.

NEUROPHYSIOLOGY AND MENTAL DISEASES

Research has contributed greatly to the clinical
advances in treating renal disease as discussed also in
another chapter. Not only do we understand normal
kidney function better but also more effective substances
have been introduced as diuretic agents for use in patient
care. Electrolyte therapy and the practical application
of dialysis in acute renal failure has developed recently.
A better understanding of the physiology of diuretic
agents has led also to improved treatment of patients with
congestive heart failure. In congestive heart failure
urinary excretion does not keep pace with salt and water
intake, even when salt intake is restricted. The relief of
edema using newly developed diuretic agents has been
a therapeutic advance of great value to both physician
and patient.

Basic to human behavior, to the excitation of muscular
contractions, to the stimulation of many secretions including even some hormones, to the activities of that
living computer, the brain, is the activity of the nerve
cells and their long extensions that are bound into
bundles called nerves and nerve tracts. A most important
achievement has been an explanation for the mechanism
by which signals are transmitted from one end of a nerve
cell down its long extension, the nerve fiber, to the other
end. Equally important is the mechanism by which

these signals called the nerve impulses are transmitted at great speed across the short space that separates the ends of two different nerve fibers. The infinitesimal signals, both electrical and metabolic, underlie sensation, movement, emotions, and consciousness itself. Physiologists have devised clever methods for recording events that take place inside the nerve cells, cells too small to be seen with the naked eye. With information about these intracellular happenings they have been able to explain the physical, chemical, and electrical events involved in the generation of these signals and their transmission along a nerve fiber and across the junction to another fiber.

There have been notable advances during the past 30 years in understanding how large numbers of nerve cells and their fibers are linked together to form electrical circuits. The general design of the brain, and the engineering principles embodied in it, are becoming apparent. We are beginning to understand how muscles are controlled and how sensations are perceived, in terms of brain circuits. Great progress has been made with regard to the nature of sleep and dreaming, a state that occupies about one-third of the human life-span.

Several of the elaborate processes that underlie vision have been elucidated. The photochemical events that occur when light strikes the retina, the electrical activity elicited in the retina by the light impulse and transmitted along the nerve pathways to the various regions of the brain, and the mechanisms by which patterns are recognized by the brain now can be described in some detail. A new technique called electroretinography, the recording and study of the electrical signals in nerve cells that constitute the retina, has been introduced recently. The procedure enables an ophthalmologist to detect hereditary progressive retinal disease long before a patient has lost his sight. Early detection offers the possibility of devising methods for retarding or preventing degenerative changes in the retina that lead to blindness.

How a signal gets across the junction called a synapse, which separates one nerve fiber from another, is a puzzling, difficult, and exceedingly important problem.

A considerable body of successful research carried out
during the past 30 years has helped to solve this puzzle.
It is now recognized that the transmission of information
in the nervous system generally involves chemical rather
than electrical relays. The transmission of impulses at
synaptic junctions results from the release of minute
amounts of chemical transmitters that diffuse across the
narrow gap separating the fibers. The award of a Nobel
prize in 1970 recognized the importance of the work. A
molecule of acetylcholine transmits the signal from one
nerve fiber to another; at least, this appears to be the
case in the main portion of the nervous system. There is,
however, another part of the nervous system called the
sympathetic nervous system. In the sympathetic nervous
system a different chemical transmitter, noradrenalin,
carries the nerve signal from one sympathetic nerve
fiber to another. After the signal has been carried across
to the other fiber, the chemical transmitter molecule
must be destroyed before it accumulates and causes
trouble. Much research, with important implications to
clinical medicine, has been directed toward understanding the inactivation process. When a compound, such
as the transmitter called noradrenalin (norepinephrine),
is broken down into component parts or perhaps changed
to another chemical form, the parts or other forms are
called metabolites. An analysis of urine for the presence
of metabolites of the transmitter noradrenalin now
constitutes a valuable diagnostic procedure in clinical
medicine. There is evidence that in the brain and in
the spinal cord amino acids may serve a transmitter
role. The participation of amino acids in transmitter
functions may prove to be a most fortuitous and invaluable discovery in investigating the central nervous
system.

Substances that affect the process of nerve signal
transmission have proved useful in anesthesiology, and in
treating hypertension, heart beat irregularities, and
psychiatric disorders. Within the past decade, using new
incisive techniques, noradrenalin and two other substances called dopamine and serotonin have been
identified in particular groups of nerve cells in the brain.

Much has been learned about the metabolism of these substances and their involvement in mental states and behavior. The findings important in neurology and psychiatry, and generally regarded as one of the most important neurological advances in the past decade, have led to the use of the chemical L-dopa in treating Parkinson's disease. Substances that relieve depression or act as tranquilizers have been discovered or developed and have helped to decrease dramatically the number of hospitalized mental patients. The findings have provided other researchers with powerful tools for investigating biochemical and physiological changes in psychiatric disorders, studies which can lead to further advances in understanding human behavior.

Research that has contributed directly to improved clinical care for cardiovascular patients will be discussed in another chapter. As so often happens in science, new information in this field has found application elsewhere. Investigators interested in the control of circulation began to study the metabolism of vasoactive agents such as noradrenalin and serotonin. A number of important new cardiovascular drugs have come from this basic research, along with new knowledge of cardiovascular control. The far-reaching benefits for patients with psychiatric disorders were quite unexpected.

As recently as 1950, half of the hospital beds in the United States were in mental institutions. At that time, half of those beds contained schizophrenic patients. Another way of emphasizing the problem is to describe the extent of schizophrenia. More than 1% of the population of the world will, at one time or another, exhibit the symptoms of schizophrenia. Of these, a large share will be disabled for longer than half their lives. In the United States alone, approximately 1 million man-years of potentially useful effort is lost each year. In monetary terms, productivity equivalent to perhaps 5 billion dollars per year is lost. To this add the anguish and the loss in productivity of the families disrupted by the psychotic, and the bill becomes truly staggering.

Schizophrenia is only one of the functional mental ailments; others include depression, mania, senile psychoses, disabling neuroses, chronic alcoholism, and mental deterioration from abuse of drugs. These neurological illnesses add an economic burden that, if accurately assessed, would appear to cost the nation as much as heart disease and cancer combined.

The development, growth and maintenance of body structure is influenced not only by complex genetic and hormonal systems, but also by the nervous system. It has been known for a long time that when nerves are damaged, muscles and other parts of the body waste away and when nerves heal other body structures regenerate. Much has been learned in recent years and we know now that nerves provide some important substance that keeps muscle and other tissues healthy. The substance and its effects on muscle and other tissues is under investigation using advanced biochemical methods. The isolation and identification of this substance, or substances, and the application of the knowledge to the treatment of degenerative processes is anticipated within a decade. The potential usefulness of such a substance in treating diseases of the central nervous system would be of immense value. When the spinal cord is severed nerve fibers do not regenerate and the crippled victim is confined to a wheelchair. A substance that would regenerate severed nerve fibers allowing them to splice would be a miraculous gift to many wheelchair victims. Although speculative it is not an impossibility. Paraplegia presently affects some 200,000 Americans at a cost to the United States in treatment and lost income exceeding 2 billion dollars annually. This toll has increased at a rate of 3,000–6,000 cases a year in the past decade because of highway accidents, sports mishaps, and battlefield injuries. While progress has been made in the areas of patient care, adjustment, and early treatment, the real hope for paraplegics lies in finding a cure.

CARDIOVASCULAR DISEASES

Diseases of the heart and circulatory system end more lives prematurely than any other ailments affecting modern man. More than 30 million people in the United

States have some form of cardiovascular disability, and over a million die each year as a consequence of such illnesses. The United States Public Health Service figures for 1967 indicate that direct costs for care of patients with cardiovascular disease was 5.1 billion dollars. Direct costs for care represents only a small part of the economic costs to the nation. How much greater would be the Gross National Product, in any year, if morbidity and premature mortality from these diseases had not interfered? The National Center for Health Statistics estimated that for 12 months beginning June 1966 cardiovascular disease caused 160 million disability days from bed disability and 452 million for restricted activity. Billions of dollars, perhaps as much as 25 billion, are lost each year in reduced productivity.

Fifteen million people or more have hypertension or high blood pressure. The sustained elevation of blood pressure leads to coronary artery disease, congestive heart failure, renal failure, cerebral hemorrhage, shock, and so on. It is well established that hypertension is more prevalent among blacks than among whites, even under similar nutritional conditions. Does socioeconomic and environmental stress cause the difference? Physiologists have demonstrated recently that hypertension can be induced in monkeys using the technique of stress. Confirmation and extension of this work may provide a better model for studying the origin and physiology of hypertension.

Essential hypertension is one of the most common potentially fatal cardiovascular diseases in this country. Prior to the late 1940's there were no effective drugs available for treating this condition, and therapy was dependent entirely on measures such as sedation, marked restriction of activities, low salt diets, and, in selected cases, relatively drastic surgical procedures such as sympathectomy, adrenalectomy, and hypophysectomy.

Several groups of drugs were introduced in the 1950's that lowered blood pressure chiefly by reducing the activity of the sympathetic nervous system at various levels. While these drugs can reduce blood pressure, and many are still used in treating certain phases of hypertension, all have definite limitations, including restricted efficacy, seriousness of side effects, and the development of tolerance to their action.

A major advance came with the introduction of α-methyldopa and reserpine, an alkaloid previously used as a tranquilizing agent. Both drugs lower blood pressure in a more satisfactory, better controlled manner by decreasing the amount of norepinephrine, the neurotransmitter in nerves involved in control of blood pressure, that is available for release. More recently, a group of drugs known as the thiazide diuretics have proved extremely valuable adjuncts to antihypertensive therapy, probably by producing directly a decrease in the tone of peripheral blood vessels in addition to their action on the kidney. At present, the blood pressure in the great majority of hypertensive patients can be controlled adequately by these drugs.

Catheterization of the heart has provided a better understanding of the cardiovascular physiology. A technique for measuring the flow of blood through the brain has given information on cerebral blood flow in disease and to refined procedures for measuring the flow in particular regions of the brain and other organs. Contractions of the heart are accompanied by electrical changes that can be recorded. For several decades this record, called an electrocardiogram, has been a powerful diagnostic tool. Only recently have we begun to understand the nature of the electrical activity of the cells whose collective effects appear in the electrocardiogram.

With microelectrodes, physiologists now penetrate the interior of single cells in the heart to determine characteristics of the electrical activity in the individual cell, and factors that may alter the activity. The work has placed the treatment of heart beat irregularities on firmer ground and new methods of treatment have been devised. Electrophysiological studies also have demonstrated a peculiar feature in the transmission of electrical impulses between upper and lower chambers of the heart. In some patients the impulse does not get through to the ventricle or lower chamber and the patient has a condition known as complete heart block. Usually a ventricular pacemaker will take over at a decreased rate, but often not reliably, and the heart may stop, leading to fainting, or syncope, or even death. Today, an artificial electronic pacemaker can be implanted under the skin and connected to the ventricle, changing the entire course of life for the patient.

COMPARATIVE PHYSIOLOGY

Primitive forms of animal life, such as the squid, the lobster, the horseshoe crab, and the jellyfish, have contributed to modern medicine by providing material for studying critical problems in cellular physiology. The modern era of excitable membrane physiology, which began with the study of the giant fibers of nerve cells in the squid, has led to much of the present knowledge about the movements of molecules carrying electrical charges. The nervous systems of many invertebrate animals like the squid contain relatively few but very large nerve cells that can be studied more readily than human nerve cells. The studies have provided important new data on intercellular interactions and other changes involved in behavioral conditioning. Much information about muscle tone and the knee-jerk reflex has come from research on the stretch receptor in the lobster. The stretch receptor is an organ in the muscle that, when stimulated, gives rise to a sudden vigorous muscle contraction. The eye of the horseshoe crab has served as a useful organ for research on vision.

A jellyfish *Aequorea* when stimulated releases a protein that glows with a cold light in the presence of calcium ions in sea water. When the bioluminescent protein is injected into single giant muscle fiber at rest, light cannot be detected. Stimulating the muscle fiber causes light to be emitted. The observation indicates that when muscle is stimulated calcium ions are released from attachment inside the muscle cell. Both the intensity of emitted light and the electrical potential of the cell membrane can be measured. The method provides a quantitative means for studying the relation between electrical potential and the release of calcium ions. While the study of lower forms of animal life may seem to have little relevance to human health and disease, comparative physiological studies have contributed often and importantly.

MACROMOLECULES AND DNA

The intellectual discipline called biochemistry relates biology to chemistry, and attempts to interpret living phenomena in terms of molecular structure and chemical reactions. The chemical reductionist approach to biological processes appears to be one of the most fruitful routes to the comprehension of biological events. To discuss major biochemical advances of the past 30 years, it is important to understand the nature of some very large molecules known as proteins and nucleic acids.

Nucleic acids, although first described about 1870, only recently have been found to be the primary structures for transmitting hereditary information. Nucleic acids were first isolated between the years 1868 and 1872 from leukocytes and from sperm cells. Prior to 1945 only a few investigators chose to face the great difficulties encountered in studying these gigantic and complex molecules.

A momentous experiment in 1944 demonstrated that the nucleic acid DNA obtained from one type of bacteria, when added to a culture of another type of bacteria, introduced a permanent change in the appearance and the virulence of the latter bacteria. Although not accepted immediately it is now known that the DNA introduced a new stable and heritable characteristic into these bacteria. A few years later DNA obtained from a virus was used to infect a bacterial cell. In this experiment complete viruses grew within the infected bacterium. The discovery of the biological activity of DNA and its recognition as the physical basis of the gene stimulated enormous interest in the nucleic acids.

Deoxyribonucleic acid consists of four essential major components, the nucleotides of the four bases called adenine, guanine, cytosine, and thymine or A, G, C, and T for short. A DNA molecule synthesized in a particular cell at a particular time is a specific substance. The molecule must contain A, G, C, and T in the right proportions. Each nucleotide used to build a DNA molecule contains a sugar molecule called deoxyribose and a molecule of phosphate in addition to one of the four bases mentioned above. A nucleotide consists of base–sugar–phosphate held together by chemical bonds to form a monomeric unit. The four monomeric units, each containing a different base, are the building blocks which link together to form a long polymer chain.

It was reported in 1950 that the ratios of A to T and G to C were essentially unity in the DNA from all organisms including viruses and man. The ratio A + T to G + C on the other hand was found to be variable from one species to another in both plants and animals. This information along with data obtained from X-ray diffraction patterns of DNA led to a proposed structure for the DNA molecule. The model suggested two long chains coiled like a spiral staircase. The backbone chain of sugar and phosphate provided the railings and the bases interspaced between formed the steps.

The helical chains embodied a feature called "complementarity." Complementarity specified that the A unit in one chain always linked with a T unit of the other chain, and a G of one chain always linked with a C in the other chain. As a consequence of this relationship, if the two chains were separated and if each strand were replicated by the formation of its complement the result would be duplication of each original chain. The model not only accounted for the base unit ratios but also provided an explanation for the exact duplication of genetic material. The feature of complementarity provided the necessary information for a chain of base units to direct the formation of a new molecule with complementary structure. The replicating mechanism preserved specific molecular structure and transmitted the structure when cells divided.

The DNA present in a single mammalian cell contains at least a billion pairs of base units. The number is quite sufficient to account for the enormous amount of information needed for directing protein synthesis and other cell activities, including inheritance.

In copulation, followed by fertilization, a sperm penetrates to the inside of an egg, carrying with it male genetic information to be put together with female genetic information already in the egg. These combined instructions are sufficient to direct the orderly and precisely timed unfolding of the manifold events leading to the growth and full development of a complete organism, whether an amoeba, a whale, or a man. Furthermore the directive information is precise: an amoeba is to be amoeba, not an insect; a whale is to be a whale, not a fish; a man is to be a man, not a monkey. What are the mechanisms by which these precise instructions are transmitted and carried out?

Proteins are the principal building materials for producing living creatures and the nucleic acids provide the blueprints and specifications. In the process of discovering these remarkable relationships biochemists and molecular biologists have reached to the molecular level of structure for an explanation of the patterns and events of life. In the realm of heredity they deal with polymers and have reduced the decisive controls to a matter of the precise order in which the monomeric units are arranged in a giant molecule.

THE PROTEINS

Proteins also are long-chain polymers, constructed from monomeric building blocks, amino acids. The amino acids are linked together by a common coupling called a peptide bound to form a long chain called a polypeptide. There are 20 different kinds of amino acids and any 1 of the 20 might be placed next to any other amino acid in the sequence in a polypeptide chain. The process of linking amino acids in sequence is anything but random, however. It is controlled by corresponding sequences in specific nucleic acids in the living cell. In the synthesis of proteins within a living cell the sequence of bases in the nucleic acids is translated into the sequence of amino acids in the protein. The discovery of the rules applying to this process of translation is one of the truly outstanding achievements of science during this century.

The proteins are the workhorses of the cell. They act as the enzymes that catalyze almost all of the chemical transformation occurring within cells. Proteins make up a large part of the structural elements of the cell wall and its internal membranes that confine and protect the watery cytoplasm and its enzymes. The collagen of connective tissues provides a framework from which to hang these cells. Proteins are the major component of protective skin and hair. They are the flexible framework of the antibodies that recognize and combat unwanted invaders. They even carry out the orders of invading viruses.

It has been estimated that a human body contains hundreds of thousands of different kinds of protein, and that as many as a trillion different kinds of protein may exist in nature. The mechanisms by which this huge number of different kinds of molecules may be made from only 20 different amino acids and each kind carry out its own specific role are now beginning to be explicable.

The nature of the forces that enable proteins to interact with one another and with other large or small molecules is of great interest. When these interactions are understood, the biochemist should be able to understand the mechanisms of cell assembly and differentiation. The answers are of critical importance in understanding the origin of cancer and in designing drugs on a rational basis.

Proteins were recognized prior to 1850 as distinct cellular components but the concept of proteins as macromolecules composed of peptide chains, with unique sequences of amino acids that can be precisely determined, has existed scarcely 20 years. The backbone, the link of atoms that hold the linear polymer together, is identical in all proteins. Variations in protein structure and function arise from differences in the short side chains, or functional groups, that are attached to every third link of the polypeptide chain. The functional groups provide the diversity that allows so many different proteins.

Crystals of a protein, hemoglobin, were observed as early as 1840 and by the year 1900 a large number of proteins could be prepared as crystals. Techniques for analyzing and determining molecular size and shape of protein molecules were nonexistent until this century, however. Beginning about 1920 new and better techniques were developed that led to an almost explosive growth of protein chemistry. The development of the ultracentrifuge in 1926 and its widespread application provided biologists with a new and remarkably incisive tool for studying biopolymers. By 1950 the size and shape of many protein molecules were reasonably well known.

In 1953, for the first time, the particular sequence of amino acids that occurs in a protein was determined. The sequence was unraveled for a rather small protein molecule, the hormone called insulin. Since 1953 the sequence of amino acids has been worked out in many other proteins, and in increasingly larger ones as techniques have improved.

Among the proteins whose sequences have been determined are the four polypeptide chains of hemoglobin. One abnormal hemoglobin, found in humans with sickle cell anemia, was found to have only one change in its amino acid sequence, the substitution of a valine residue for a glutamic acid residue in the sixth position of two of the four polypeptide chains. The mutation can arise from substitution of an adenine (A) for a thymine (T) in the DNA directing synthesis of the messenger RNA that codes for the hemoglobin polypeptide in which the change is found. About 150 such mutations in human hemoglobin have now been found and characterized.

The properties of proteins are determined by the order in which specific amino acids are inserted into the linear polypeptide chain. Proteins, however, do not exist as straight chains of atoms in living cells. They form helices, sheets, or completely folded structures depending on the properties of the backbone and of the attached functional groups.

The analysis of X-ray diffraction patterns of proteins led to our present understanding of the three-dimensional structures of protein. A beam of X rays impinging on a crystal will be diffracted by the atoms in the crystal in proportion to the mass of the atoms encountered. With atoms arranged in a regular pattern, the beam will emerge from the crystal in a regular pattern that can be recorded on film or by scanning detectors. The three-dimensional structure of the molecule can then be calculated from information in this pattern if the sequence of amino acids is also known.

ENZYMES

The synthesis and breakdown of every kind of molecule essential for life are the result of enzymes. All enzymes are proteins. Enzymes are required to synthesize and digest starch, sugars, fats, nucleic acids, proteins, and innumerable other substances. Each enzyme catalyzes a specific step in the frequently very elaborate processes of synthesis and decomposition. In a living organism hundreds of different reactions are proceeding simultaneously, each catalyzed by its own specific enzyme. Each enzyme is specifically adapted to catalyze a particular chemical reaction or a small class of related actions.

The smaller building blocks that are used to make macromolecules are also the products of elaborate and sequential enzymic processes. The amino acids used in building protein molecules and the nucleotides used in building nucleic acids must also be synthesized. Clearly such complicated and sequential events must be carefully regulated if an organism is to function at all, let alone efficiently. If any particular product is being manufactured too rapidly relative to others, it will accumulate and be present in a cell in wasteful and perhaps even harmful excess. Regulation is achieved in various ways and our understanding of the structural basis for these remarkable events is still very limited. Scarcely a dozen years have passed since the existence of regulatory proteins was clearly recognized. Enzymes in increasing numbers are being discovered which can be switched on, off, and into other modes of activity, in response to specific triggering ligands.

Prior to 1950 biological processes were thought to be controlled either by hormones or by the nervous system. The unraveling of the pathways of metabolism opened up new territory for exploration. The elucidation of pathways revealed the existence of many new enzymes whose properties and mechanism of action required clarification. One of the primary salients of biochemistry, an arena where intermediary metabolism, enzymology, and molecular mechanisms conjoin, is the matter of regulation of enzyme activity. A detailed understanding of the regulation of metabolism promises to reveal some of the innermost secrets of living matter. These regulatory mechanisms relate to one of the most significant features of living things, the ability to adapt to changing circumstances.

A development having a profound influence on biology, and therefore on man's understanding of man, was the recognition that there are two kinds of information transfer in living cells and that the two kinds are interrelated. One is the transfer of information through DNA that controls heredity and reproduction and ensures an exact duplication of cells and organisms. The second is a transfer of information from DNA to RNA to protein, permitting and regulating thousands of chemical reactions that comprise the totality of cell growth, cell maintenance, and cell activity.

The DNA of a cell begets the DNA in the two cells formed when the original cell divides. The DNA in any cell determines and directs the function of the cell by ultimately preordaining the nature and variety of the enzymes of the cell and the order in which, during cell development, the various patterns of enzymes are formed. The enzymes as indicated facilitate and speed up thousands of distinct but coordinated chemical reactions in the cell. The DNA specifies which of many thousands of enzymes should be present in a particular cell. It does this by acting through an intermediary, another kind of nucleic acid called RNA. It is the RNA that controls directly the creation of specific proteins needed by the cell.

BIOSYNTHESIS

Enzymes have been discovered and purified that cause unattached nucleotide units to link together into chains the sequence of bases being directed by a DNA molecule called a primer. In the new chain, growing in length along the template chain, the base units are always complementary to the base units of the pre-existing template chain, in accordance with the principle of complementarity mentioned above. The enzyme that facilitates this chain production is DNA polymerase because the product, a polymer, is constructed from monomer units, the nucleotide bases. Other investigators soon discovered RNA polymerases, enzymes that would build RNA chains using a DNA chain as the template. The RNA chains that are formed act as "messengers" for protein synthesis. The chains are called "messengers" because the RNA carries a message copied from preformed DNA. The message is carried from the nucleus into the cytoplasm of the cells, where the apparatus for protein synthesis occurs. The "message" carried on the RNA molecule specifies the sequence of amino acids in protein molecules.

It is obvious that the message is coded in a particular sequence of nucleotides. When DNA replicates, base unit A on the template chain will pair with base unit T in a newly formed chain. When DNA provides the template for a messenger RNA chain, base A pairs with base unit U. The letter U stands for uracil, a pyrimidine base found in RNA. The messenger RNA directs the formation of protein by a second step in which the message is translated to a linear sequence of amino acids in a protein. Three base units of the messenger RNA sequence specify at each point which amino acid is to be inserted into the protein molecule. This understanding of the genetic code was elaborated during the past decade. Recently discovered elements of the code are the "punctuation marks" that indicate when the construction of a protein chain is to be initiated and terminated.

Our knowledge of the mechanisms by which proteins are synthesized in the cell has been acquired almost entirely from research in the last 2 decades. All of the important steps in the process were first elucidated with bacterial systems. While an early association of cell growth with increased RNA content implicated the latter in protein synthesis, the involvement of nucleic

acids in protein synthesis has been clarified only by
detailed biochemical studies of simplified, reconstructed
cell-free systems. Research directed toward the problem
of how living cells manufacture proteins has become a
major activity of biochemists.

As increasing numbers of investigators examined the
synthesis of proteins in bacteria, yeast, plant, and animal
cells it became clear that the basic steps by which the
living cell makes protein are the same in all cells, which
added validity to one of the fundamental precepts of
modern biology. The process requires "messenger RNA,"
a form of RNA called "transfer RNA," and extremely
tiny cell particles called "ribosomes." The ribosomes,
hardly larger than genes, contain protein and nearly all
of the RNA in the cell. In the synthesis of a protein the
messenger RNA attaches itself to the ribosome. The
ribosome serves as a scaffolding on which the messenger
RNA carries out its function. The amino acids from
which the protein molecule is to be constructed are
grasped, so to speak, by appropriate and specific transfer
RNA molecules and brought to the ribosome scaffold
where the messenger RNA specifies the sequence into
which the amino acid molecules are linked together to
form the protein chain. As amino acids are added the
protein chain elongates, forming the specified protein.

The process can be compared to the construction of a
brick wall where the hod carrier, or transfer RNA,
selects the proper bricks, or amino acid molecules, and
carries them to the bricklayer, or messenger RNA.
From his ribosome platform, the bricklayer fits the bricks
together in an appropriate order. The hod carrier and
the bricklayer keep things going through a communication system, the code. But this is not all. To keep the
building process going sufficiently fast, the hod carrier
"transfer RNA" and the bricklayer "messenger RNA"
are speeded up by some efficiency experts, specific
cellular enzymes that "catalyze" the reaction. Within
the cell individual enzymes catalyze each step from the
initial specific linking up of each amino acid to a molecule of transfer RNA, through the successive additions of
amino acid molecules to the growing protein chain, to

the release of the finished protein. This rapid and marvelous process has come to be understood in part, but by no means in every detail. The enzymes involved in protein syntheses are only a few of the multitude of different enzymes that help to keep the machinery of life moving.

Although enzyme technology began in antiquity with the fermenting of beer and wine, it was not until the 1920's that enzymes were shown to be proteins. An explosive development in enzymology has occurred during the past 2 decades, in part at least because of improved isolation and purification methods. The development of chromatographic methods was certainly a major factor in facilitating the separation and purification· of enzymes. Enzymes are catalysts that affect the rate of chemical reactions. The enzyme itself is not altered permanently in the process. The study of biochemical reactions has grown increasingly important in the past 15 years.

It has been found that inhibitors exist for many enzyme reactions and that the inhibitors can be used as a pharmacological tool as well as for clinical treatment of disease. As our understanding of the mechanism of enzyme action improved, a new function of enzymes was revealed. Certain key enzymes at particular controlling points in the processes of metabolism act as metabolic regulators. Recognition of the regulator function improves our understanding of the relationship between structure and function in enzymes and how metabolic processes in the cell are self-regulated. It has been found, for example, that the action of some enzymes is inhibited by products of the reaction. The finding suggested that parts of the enzyme molecule may be changed in ways that alter the speed of the chemical reaction it is influencing. Many enzymes have been found to exist in multiple forms, called isoenzymes. The isoenzymes are particularly interesting substances that are chemically and physically distinguishable but nevertheless catalyze the same chemical reaction. Although the significance of isoenzymes is not yet well understood, it has been found that abnormally high levels of a particular isoenzyme in the blood may indicate a disease process in a particular tissue, a useful contribution to diagnostic medicine.

METABOLIC EVENTS

Metabolism is the work that enzymes carry out: the conversion of food into the compounds required for maintenance and growth, including synthesis of all the nucleic acids, proteins, and cell wall components. Simple life forms, such as the bacteria, use very few different nutrients. *Escherichia coli*, a bacterium found in the intestinal tract, can grow and multiply when supplied with only glucose and a few salts. Man has very complex dietary requirements, including many different amino acids, vitamins, certain types of fats, and mineral salts. The bacterium, therefore, is capable of more synthetic reactions than is man. It has been estimated that *E. coli* contains about 1,000 different kinds of small molecules. These are either involved in conversion and transfer of chemical energy or are used as, or are intermediate in formation of, the building blocks for the 4,000 large molecules, the nucleic acids, proteins, and polysaccharides that are the structural and functional framework of the *E. coli* cell.

At the time biochemical studies were initiated, early in the century, an understanding of how cells burn (oxidize) sugars, fats, and amino acids was crude and fragmentary. The mechanisms by which simple sugars, fats, vitamins, and substances called purines and pyrimidines are utilized in biochemical processes were slowly elucidated. Many of the major metabolites of the cell had been isolated by 1940. Almost all of the amino acids, nucleotides, sugars, lipids, and vitamins that are now known had been identified and their chemical structures elucidated. Some of the major pathways of metabolism were partially characterized by 1940. It was known, for example, that the six-carbon chain of glucose was broken to form two two-carbon molecules of ethanol and two molecules of carbon dioxide in fermentation by yeast. A compound, now known as "cyclic AMP," discovered in the 1950's has been found to be the agent through which many hormones produce their effects, for example, in the metabolism of carbohydrates. The steroid hormones appear to act through a different mechanism, however. After their discovery the chemical role of vitamins and other accessory nutritional factors was gradually worked out during the period 1940 to 1960. We have now developed detailed knowledge of how the cell synthesizes large polymers such as proteins, starch, glycogens, and nucleic acids.

ENERGY

Photosynthesis, the most important and fundamental of all energy mechanisms, is a process in which the radiant energy of the sun is captured and stored in the form of organic molecules. Photosynthesis occurs in the green plants that envelop the land and live in the oceans. The energy stored in the green plants provides the primary source of energy for all animal life on the planet. The ultimate source of all energy in living and fossil structures is the sun.

An extremely important compound called adenosine triphosphate (ATP) was first isolated in 1929 from natural material. The ATP molecule represents a kind of universal fuel used to provide energy in biochemical reactions. Its pivotal function in all cells has been established by a multitude of studies carried out since 1940. Synthesis of the molecule in 1948 confirmed its molecular structure. The chief source of energy available to cells that consume oxygen is the oxidation of hydrogen atoms in the molecules of organic compounds. The oxidation reaction takes place mainly in particles called mitochondria, located in the cytoplasm of cells. Mitochondria, much larger than ribosomes, another intracellular structure already mentioned, are miniature power plants that provide energy for the cell. Adenosine triphosphate is formed in the mitochondria and used for all the cellular processes that require energy, processes such as muscle contraction, the conduction of nerve impulses along a nerve fiber, the maintenance concentration gradients, and a host of other intracellular chemical reactions. Most of our present knowledge about the utilization of energy in the cell has been attained since the central role of ATP was recognized in 1940.

THE MOLECULAR BASIS OF DISEASE

There are many reasons for the great advances in the biomedical sciences during the past quarter century other than substantial financial support of research. A reason of great importance to biology and medicine is the concept that biological structures are organized on a molecular basis and that diseases originate in mal-functioning molecular mechanisms. This reductionist viewpoint gained acceptance slowly during the second quarter of the 20th century.

In numerous instances an investigation of the molecular basis of a disease has led to a precise definition of the cause of the disease, and to a detailed understanding of the mechanisms by which the disease progresses. An outstanding example exists for sickle cell anemia. This disease is genetic in origin and results from an error built into the protein-synthesizing process. As mentioned previously, the amino acid sequence of a protein is encoded in the sequence of nucleotides in a particular stretch of chromosomal DNA, each amino acid being specified by a triplet of nucleotides.

A mistake in the message on a DNA molecule produces a consistent mistake in an amino acid appearing at that point in a particular protein. In the case of sickle cell disease an incorrect amino acid in a protein chain in the hemoglobin molecule results in a defective molecule. The ability of the molecule to perform its function of trans-porting oxygen from the lungs to the cells where oxygen is consumed is impaired by the defect.

Since the discovery of the structural defect in the hemoglobin of people with sickle cell anemia, other abnormal hemoglobin molecules have been identified in the blood in individuals suffering from other blood diseases. In many of these variant hemoglobins the abnormality consists of a simple substitution of one amino acid for another. More than 150 slightly different human hemoglobins have been found and characterized.

Such knowledge encourages us to believe that we are beginning to understand biological processes in molecular terms.

Among advances of the future that may be anticipated are:

1) Chemical synthesis of familiar enzymes with the introduction of new enzymatic properties by chemical modification, analogues to the many drugs that occur naturally and that have been modified chemically to obtain greater specificity in action.

2) Development of methods for introducing genetic information into cells, with the hope of treating diseases of genetic origin. Genetic disorders offer a unique opportunity to combine the concepts of genetics with the tools of biochemistry in the study of metabolic processes in humans. Such studies have already proved remarkably successful with microorganisms. The lessons learned in these studies make it quite clear that genes contain the code that specifies the structure of protein molecules. Another fundamental lesson, now well established, tells us that a mutation in a gene, that is a chemical change in a purine or pyrimidine subunit found in the DNA molecule, will result in an alteration in the specific protein for which a particular gene carries the code. The result may be harmless or disastrous depending on the importance of the protein in the metabolic pathway and the effect of the structural change on the function of the protein. When a disease has been identified as genetic in origin, for example the disease called phenylketonuria, the task of the biochemist is to identify the altered protein which is specific to that disorder. Success can lead to practical applications in managing the disease and also to a better understanding of normal metabolic processes. Malfunctioning at one point in a metabolic pathway, for example, may affect events far removed from the primary defect and give rise to bizarre clinical manifestations seemingly unrelated to the basic lesion. More than one hundred diseases have now been recognized as genetic in origin resulting from inborn errors in metabolism.

3) An increased understanding of the chemical factors that influence growth, differentiation, and aging of cells and tissues.

4) Expanding the range and effectiveness of drugs used in human disorders.

5) A better chemical–immunological basis for reliable tissue transplantation techniques and for controlling or preventing self-immunization to factors in one's own tissues.

6) Advances in knowledge of the chemistry of brain functions involved in such things as thinking, feeling, and behavior.

Bacteria, and the viruses which infect bacteria, have provided remarkably fruitful model systems for understanding some of the biological processes occurring in humans. The concept that many or most of the individual steps in the metabolism of a compound are either identical or similar in the most diverse forms of life is of inestimable importance in biomedical research. It has enabled vast numbers of experiments to be carried out cheaply, rapidly, and under carefully controlled conditions using bacteria, viruses, plants, or small animals. Only a small fraction of the experiments could have been undertaken with human subjects. The enormous savings in time, in labor, and in dollars is beyond calculation. The results of such studies not only increase our faith in the concept but permit physicians to treat patients with confidence undreamed of a generation ago.

RATIONAL DESIGN OF DRUGS

An area of biology allied to biochemistry and in which biochemistry is deeply involved is pharmacology. An increase in the support of pharmacological research, additional trained manpower, advances in other areas of science useful in the development of pharmacology, and new sophisticated methods of analysis have all contributed to the rapid development of pharmacology. Pharmacologists are now beginning to elucidate the mechanism of reactions between drugs and elusive cell components on which the drugs exert pharmacological or therapeutic effects. A special subdiscipline, clinical pharmacology, has arisen that relates therapeutic values and toxic potentialities of new drugs more objectively and frequently in molecular terms. Pharmacological studies, previously limited to laboratory animals, can now be carried out safely in humans. Such studies determine drug uptake, the biochemical mechanisms of drug action, and the degradation and excretion of drugs. Outstanding pharmacological advances have occurred in the areas of infectious diseases, cardiovascular diseases, and endocrinology, as discussed elsewhere.

During and after World War II both laboratory animal experiments and clinical studies with substances called nitrogen mustards, originally developed as chemical warfare agents, have led to a large group of drugs which, although they do not cure cancer, can prolong a useful and relatively pain-free life for many cancer victims. Another group of drugs interrupts cell division by interfering with the duplication of DNA, a process which occurs in a run-away fashion in cancer. The mode of action of both groups of drugs explains two important common features: 1) their toxic side effects on tissues, such as bone marrow, where cell division occurs most actively, and 2) their usefulness also as immunosuppressive drugs. Immunosuppression is discussed in another chapter.

It will be pointed out elsewhere that folic acid has a beneficial effect in childhood leukemia. Recently pharmacologists have developed additional clinically useful drugs for the same purpose. In recent years a miscellaneous assortment of pharmacological agents has been introduced for the treatment of cancer. The disparate results in the chemical treatment of bacterial infections and that of malignant disease are painfully evident. Bacterial infections are almost always cured whereas malignancies are benefited only temporarily. It seems unlikely that drugs that can cure cancer will be developed until more is known about the differences between the metabolic requirements of normal and malignant cells. When and if drugs can be developed that will

interfere with the metabolism of malignant cells without
altering the metabolism of normal cells, a chemical cure
for cancer may be forthcoming.

It is difficult to design useful drugs against viruses.
Unlike bacteria, viruses multiply only within the cells
of the organism they infect. Some drugs have been
developed, nevertheless, that are effective at certain
stages of viral infection, and have been used successfully,
for example, in certain eye infections. In any case a
large amount of difficult and laborious research will be
required to find answers to these questions.

Nowhere has progress been more dramatic in recent
years than in the area of severe mental illness. The factors
most responsible for progress have been the development
of effective tranquilizer and antidepressant drugs. While
there is general agreement that these drugs may not cure
mental disease, they are beneficial in controlling the
symptoms sufficiently to permit patients to function
effectively outside a hospital. Many patients, although
still under treatment, are able to hold jobs and enjoy
useful lives.

APPLICATION OF BASIC KNOWLEDGE

The history of the poliomelitis vaccines is an example
of success in applying basic concepts to a specific
problem. The story begins in 1908 when research showed
that poliomyelitis could be transmitted to monkeys, a
discovery that gave investigators an animal to use for
experimental study of the disease. The discovery that
poliomyelitis was caused by a virus was followed by
finding that strains of poliomyelitis virus could be classified into three types. Another significant advance came
with the discovery that poliomyelitis virus could be grown
on tissue in test tubes, a finding that enabled other
investigators to develop methods for producing poliomyelitis virus for vaccine. The development of both Salk
and Sabine vaccines carried these and other indispensable basic discoveries to pragmatic fruition.

The science of microbiology interfaces with many areas
of medicine. Whereas research in pathology considers

the course of events after the disease exists, research on infectious disease emphasizes causation. The continuing accumulation of evidence that viruses may cause certain forms of cancer, chronic disorders such as arthritis, and degenerative diseases of the nervous system provides examples of emphasis on etiology. The broad field of immunology, with its ramifications in many medical

TABLE 1. *Death decline*

	% of Decline
Polio, because of availability of vaccines	100
Whooping cough, because of availability of vaccine	100
Tuberculosis, because of availability of antibiotics and other drugs	51
Dysentery, because of availability of antibiotics	50
Asthma, because of availability of cortisone	35
Meningitis, because of availability of antibiotics	31
Syphilis, because of availability of antibiotics	33
Nephritis and nephrosis, because of availability of antibiotics	27

TABLE 2. *Estimated medical care costs for poliomyelitis avoided, 1955–1961*

	Inpatient	Other	Total
Deaths avoided 12,464 at $20/day for 15 days	$3,739[a]	n.a.	$3,739
Severe disability avoided 36,352 at $20/day for 120 days plus $2,000 in other costs	87,245	$72,704	159,949
Moderate disability avoided 58,133 at $20/day for 60 days plus $1,000 in other costs	69,760	58,133	127,893
Slight disability avoided 32,743 at $20/day for 30 days plus $300 in other costs	19,646	9,823	29,469
No disability avoided 14,307 at $20/day for 15 days plus $100 in other costs	4,292	1,431	5,723
Total	184,682	142,091	326,773

Calculations by the National Foundation. [a] In thousands of dollars.

areas, such as prenatal and infant mortality, organ transplantation, heart disease, arthritis and rheumatic fever, and blood disorders, is a relatively recent offspring of research with microorganisms. The routine use of uncontaminated tissue cultures and germfree animals was made possible by the same antibiotics that enable physicians to control and, in some instances, virtually eliminate formerly fetal infectious diseases such as diphtheria, cholera, typhoid, endocarditis, bacterial pneumonia, and tuberculosis. The great strides in surgery would have been impossible without present techniques for asepsis and control of infection.

Research progress in the development and improvement in antibiotics and immunological prevention and control of infectious diseases has achieved striking declines in death rates from disease where these treatments are effective (Table 1).

Statistics for poliomyelitis, measles, rubella, and tuberculosis provide an example of the impact of microbiological research on human disease.

Poliomyelitis. A most dramatic decline in poliomyelitis came with the development of polio vaccine. Contrasted with 57,879 polio cases reported in 1952 only 41 cases of acute poliomyelitis were reported in 1967, the lowest number of cases reported in the United States since reporting began in 1912. The elimination of the disease reduced the costs of health care dramatically, as shown in Table 2.

Measles. A vaccine for measles, a common disease that once killed 500 children each year and left others with lasting handicaps, including hearing disorders and mental retardation, became available in the spring of 1963. More than 11 million children have now been vaccinated under a nationwide, federally supported vaccination program.

Rubella. It is estimated that 20,000 infants in the United States suffered congenital defects from the 1964 epidemic of German measles. An almost equal number of fetus deaths were recorded. A new vaccine was developed in 1969 and is being used on a national basis. According to the Center for Disease Control of the

Communicable Disease Center approximately 13 million children have been immunized since the vaccine became available. The CDC reports that an additional 3 million children have received the vaccine in private physicians' offices.

Tuberculosis. Some 85,607 cases of tuberculosis were reported in the United States in 1952. An estimated 45,546 active cases were reported in 1967—a decline of 47%. With the distribution of streptomycin which began in 1946, the death rate from tuberculosis has declined 91%.

TABLE 3. *Estimated economic costs due to measles and benefits due to immunization, United States, 1963–1968*

Costs	Without Immunization	With Immunization	Benefits Due to Immunization
Direct, medical			
Total	$673,990ᵃ	$395,592	$278,398
Physician services in office			
Encephalitis cases	402	229	173
Other acute cases	86,701	49,442	37,259
Physician services in hospital			
Encephalitis cases	4,523	2,552	1,971
Other acute cases	6,892	4,137	2,755
Hospital services			
Encephalitis cases	26,334	13,975	12,359
Other acute cases	43,927	26,437	17,490
Gamma globulin for contacts	5,558		5,558
Lifetime care for mentally re-tarded	499,653	298,820	200,833
Indirect, loss of productivity			
Total	604,667	351,522	253,145
Premature death	105,944	56,040	49,904
Mental retardation	407,753	241,320	166,433
Work losses	90,970	54,162	36,808
Grand total	1,278,657	747,114	531,543

ᵃ In thousands of dollars.

TABLE 4. *Institutions and beds set aside for the care of tuberculosis patients in specified years*

Year	No. of Institutions	Federal and Nonfederal Beds for TB Patients
1942	710	100,275
1948	636	102,119
1951	694	106,087
1954	654	113,855
1955	650	108,235
1957	546	94,031
1959	497	87,270
1961	432	67,634
1963	409	60,363
1965	387	52,781
1967	348	43,069
1969	296	32,822

Source: Tuberculosis directories published by NTRDA and PHS.

With the exception of an antibiotic treatment of infectious diseases, almost all drugs merely relieve symptoms rather than cure the disease itself. Ignorance about two things is responsible for this state of affairs. We are ignorant of the cause or causes of the disease and we do not understand mechanisms of drug action necessary to cure the disease. The momentous achievements of the last 30 years in experimental biology lead us to expect that this situation will change in the next several decades. Sufficient support of basic research on the underlying and essential processes of disease and on the molecular mode of action of pharmacological agents offers great opportunities to alter the present state of affairs.

The deficiencies in modern medical treatment are most conspicuous when physicians are hampered by incomplete information and inadequate diagnostic tools and are required to use palliative procedures because they lack a clear understanding of disease mechanisms. A higher priority for support of basic research in biology offers the most inexpensive and most expedient route to better health care. For disease processes that are understood and for diseases which can be prevented or rapidly diagnosed and cured, the cost of health care is a small fraction of the costs of care and treatment of diseases about which we are ignorant.

SELECTED ADDITIONAL READING

BITTAR, E. E., AND N. BITTAR. *The Biological Basis of Medicine.* New York: Academic, 1968, vols. 1–6.

GOODMAN, L. S., AND A. GILMAN. *The Pharmacological Basis of Therapeutics* (4th ed.). New York: Macmillan, 1970.

GREEN, D. E., AND R. F. GOLDBERGER. *Molecular Insights into Living Processes.* New York: Academic, 1966.

KENDREW, J. C. *The Thread of Life: An Introduction to Molecular Biology.* Cambridge: Harvard Univ. Press, 1966.

LARNER, J. *Intermediary Metabolism and Its Regulation.* Englewood Cliffs, N. J.: Prentice-Hall, 1971.

LEHNINGER, A. L. *Biochemistry.* New York: F. A. Worth, 1970.

SCHMITT, F. O. *Macromolecular Specificity and Biological Memory.* Cambridge, Mass.: MIT Press, 1962.

TAYLOR, J. H. *Selected Papers on Molecular Genetics.* New York: Academic, 1965.

WATSON, J. D. *Molecular Biology of the Gene* (2nd ed.). New York: Benjamin, 1970.

WOLD, F. *Macromolecules: Structure and Function.* Englewood Cliffs, N. J.: Prentice-Hall, 1971.

YOST, H. T. *Cellular Physiology.* Englewood Cliffs, N. J.: Prentice-Hall, 1972.

Chapter III

Clinical medicine

GEORGE E. BURCH, M.D.

THIS CHAPTER SUMMARIZES some of the most important advances of the past 30 years in knowledge of diseases and in methods for maintaining health and caring for the sick. Apart from accidents, such as broken legs and body burns, a person becomes ill when an infectious or toxic agent from outside attacks his body or when internal mechanisms go wrong, as in aging or in certain inherited diseases.

The function of the physician is to help people either to overcome disease or, better still, to avoid getting sick, and to minimize or repair damage caused by disease. To do this successfully, he must have an effective arsenal of medicine and treatment techniques for prevention, cure, and repair. Today, though he has these as never before, methods for prevention and treatment of major diseases are still poorly effective or completely lacking.

Although medical knowledge and practice have improved at a continuously increasing rate during the last 200 years, advances in the past 30 years have been astonishing. Because of the great increase in medical knowledge, it has become impossible for one person to master the whole, and there has been an inevitable specialization. And so today, in addition to the general practitioner, there are many medical specialists to whom the general practitioner refers his patients for special diagnosis or treatment. Some areas, such as internal medicine and surgery, have become further split into subspecialties for the better care of the sick.

55

CHAPTER III   The infectious diseases are caused by bacteria and viruses. It is true that some infectious diseases showed a decline in morbidity and mortality before 1940, the rate of decline of which was not appreciably modified by the newer drugs available. However, the newer drugs themselves created problems and there developed an increasing incidence of diseases and disorders produced by the newer drugs and medical treatments (the so-called "iatrogenic diseases"). Furthermore, a certain amount of backsliding has been taking place recently in, for example, venereal disease and hospital infections, which by some are thought to be approaching "epidemic"

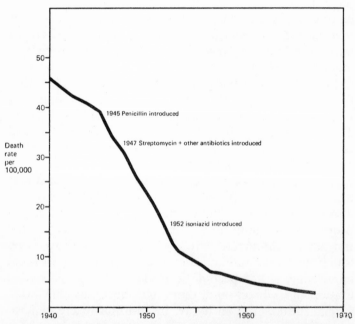

FIG. 1. Age adjusted death rate per 100,000 population from all forms of tuberculosis since 1940. *Source:* Office of Health Statistics Analysis, National Center for Health Statistics, Health Services and Mental Health Administration, Public Health Service, Department of Health, Education, and Welfare (1971).

proportions again. Perhaps this trend is associated with such factors as the widespread availability of the "pill" or an increased resistance of bacteria and other pathogenic organisms to the drugs. These are some of the human hazards that mankind is creating for himself in his quest for advances in human welfare.

Prior to 1940 many infectious diseases stalked the country: syphilis, tuberculosis, pneumonia, dysentery, meningitis, influenza, childbed fever, and others. Special tuberculosis sanitaria and hospitals, filled with debilitated and dying people, dotted this and other countries (Fig. 1). Syphilis strode triumphantly over the bodies of its victims. Physicians were unequipped to combat the scourge of these diseases, even though in 1928 a powerful weapon, penicillin, had been discovered. For 12 years this discovery, of tremendous potential importance for human health, lay unregarded and unused. Then, stimulated by World War II, researchers in England found that penicillin could cure infections in animals and man and, in this country, government and industry initiated a program of mass production of penicillin, the first antibiotic to be used extensively in clinical medicine. Since then many more antibiotics have been discovered, making it possible for physicians to treat a great variety of bacterial infections. This and an earlier finding that sulfa-containing drugs, such as sulfanilamide, were also effective against bacterial infections, led to an avalanche of new drugs during and after the war. For the first time an army was not afflicted by epidemics of contagious diseases or high mortality due to infected wounds. Tuberculosis was brought under control and sanitaria were closed or turned to other uses. Syphilis, a major threat to human health for centuries, rapidly decreased; in large city clinics the number of patient visits fell from thousands to dozens per year. New and improved vaccines were used to prevent severe viral infections such as yellow fever, and specific drugs were used against systemic fungus infections. Tetanus toxoid, known before 1940, was found to be enormously effective in the prevention of tetanus, one of the major hazards of war and accidental injuries.

TABLE 1. *Death rates per 100,000 population from selected causes of death in the past 4 decades*

Cause of Death	1940	1950	1960	1968
Selected infectious diseases				
Influenza and pneumonia[a]	70.3	31.8	37.3	34.9
Measles	1.7[b]	0.3	0.2	A[c]
Poliomyelitis	0.8	1.3	0.1	A[c]
Syphilis and its sequelae		5.0	1.6	0.4
Tuberculosis		22.5	6.1	3.3
Selected chronic diseases				
All cardiovascular diseases	407	494.4	515.0	511.0
Heart diseases		356.8	369.0	372.9
Cerebrovascular disorders		104.0	108.0	104.8
Malignant neoplasms	120.3	139.8	149.2	159.6
Diabetes mellitus	26.6	16.2	16.7	19.2
Cirrhosis of the liver		9.2	11.3	14.5
Early mortality in infancy	39.2	40.5	37.4	21.2
Accidents	73.2	60.6	52.3	55.8
Homicides and suicides	20.7	16.7	15.3	17.8

Source: Office of Health Statistics Analysis, National Center for Health Statistics, Health Services and Mental Health Administration, Public Health Service, Department of Health, Education, and Welfare (1971).   [a] Excluding pneumonia of newborn.   [b] 1941 data.   [c] A = less than 0.05.

The continued effect of these improved weapons against infectious diseases was dramatic (Table 1 and Fig. 2). Patients with bacterial meningitis, when properly treated, seldom died. Congenital syphilis and syphilitic heart disease became rare. Pneumonia patients quickly recovered. Cases of classical rheumatic fever greatly decreased. The long course of typhoid fever, undulant fever, and other such diseases was shortened to a few days. High fevers associated with blood poisoning, abdominal infection, and infection of the heart were brought under control. Fevers of more than 2 or 3 days became of special concern. Postsurgical infections were drastically reduced; infections of the ears and mastoids, previously such a threat to children, were seldom seen; and persons with many chronic diseases could now expect a longer life. One of the greatest international achievements in the area of blindness was the control of trachoma by certain antibiotics and the first isolation of the agent of this disease in 1957. Antiviral drugs began to appear for use against smallpox, influenza, and local infections. An especially important discovery was that viruses cause

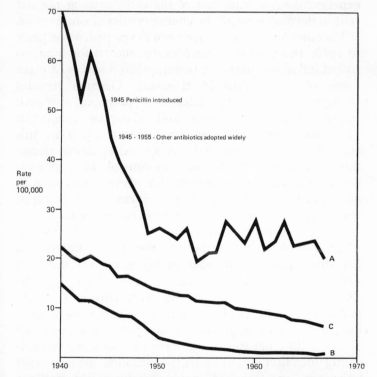

FIG. 2. Age adjusted death rates per 100,000 population for: *A*) influenza and pneumonia, *B*) syphilis, and *C*) rheumatic fever and rheumatic heart disease since 1940. *Source:* Office of Health Statistics Analysis, National Center for Health Statistics, Health Services and Mental Health Administration, Public Health Service, Department of Health, Education, and Welfare (1971).

tissue cells to produce a substance called interferon, a protein that is effective against viral infections; this discovery has paved the way for further antiviral developments.

Certain diseases, previously called "degenerative diseases," are now recognized as caused by viruses, the character of which is not yet well known. The recognition of analogous diseases in lower animals such as sheep has provided models for further study of these "slow" viruses. Certain viruses, known to cause tumors in lower animals, can induce malignant transformation, in the laboratory, in cells of human origin, a process that involves the incorporation of some of the viral genetic material with that of the cells, causing normal cells to develop some of the characteristics of cancer cells.

Vaccines and toxoids have been developed, some prior to 1940, and used with considerable success for the control of influenza, paralytic poliomyelitis, measles (a main cause of encephalitis in children), German measles (dangerous to unborn children of pregnant women), rabies, tetanus, diphtheria, and whooping cough. In 1957 and 1968, influenza epidemics pending in this country were detected early by special epidemic detection centers and efforts made to control them. A contaminating virus of monkeys, that causes infections in man and tumors in animals, was discovered, leading to better vaccines and contributing to cancer research.

These new drugs and vaccines have had important effects not only with respect to specific infections and parasitic diseases, but also in other disease areas in which infectious agents cause complications, such as the fields of cardiovascular and other chronic diseases, pediatrics, and geriatrics. The new drugs have prolonged the lives of many persons with cardiovascular disease who otherwise would have died of some complicating infection. In the past, pneumonia, urinary tract infections, and other diseases actually killed patients afflicted with serious heart disease; to some extent they still do. Many of the diseases that formerly afflicted, crippled, or killed children have been virtually eliminated or become so infrequent that they are no longer a

serious threat. The availability of the antibiotic streptomycin, and the synthetic drug isoniazid, has almost
eradicated tuberculosis in the United States and greatly
reduced the incidence of this disease in other parts of
the world. The mortality from other pulmonary disorders has decreased because of antibiotics and other
drugs. Antibiotics and other anti-infectious drugs and
improved sterile techniques have decreased hospital infections and increased aseptic conditions in hospitals
and operating rooms. Because complicating infections
can now be controlled, modern cardiac surgery and
other extensive operations are now feasible.

PARASITIC DISEASES

Malaria remains a major disease problem in the warm
regions of South America, Africa, and Asia. A major
accomplishment in the past 30 years has been the
elimination of malaria as an endemic disease in the
United States. The successful eradication programs in
other advanced countries result from the use of improved
control measures learned from World War II. An increased knowledge of the nature of the parasite, the use
of more effective insecticides, and greatly improved
drugs that became available in 1946 and thereafter
have proved highly effective in bringing malaria under
control.

In 1960 it was found that some malarial parasites are
host-interchangeable, being able to inhabit either man
or monkey, a discovery important in improving the
control of malaria.

It is now known that a high proportion of parasitic
diseases are transmitted from wild and domestic animals
to man instead of, as previously thought, mainly from
man to man. A chronic debilitating disease in humans
(toxocariasis) is caused by an intestinal worm common
in dogs and cats. Another disease caused by a lungworm
of rats (angiostrongyliasis) is spread to man by rats and
land snails and causes an acute meningoencephalitis in
humans. A disease first recognized in 1960 in Japan and
Europe (anisakiasis) is caused by worms from uncooked

marine fish that invade the human stomach and intestine. The revised concept of the relationship between parasites and man has led to improved public health measures.

A new parasitic infection caused by amoebae found in soil and stagnant water was discovered in 1965. The amoebae pass through the membranes in the nose to the brain causing a disease which is usually fatal within a week.

New knowledge about the transmission as well as improved diagnostic and treatment methods have brought under control another often rapidly fatal disease (toxoplasmosis) that is caused by a protozoan parasite in cats and other animals.

COMMUNITY HEALTH AND EPIDEMIOLOGY

Infectious and parasitic diseases are of great importance to the public health officer concerned with the delivery of health services to the community. The objective is to prevent, or stop the spread of, and wipe out disease. A public health officer must know the sources of infection, the paths they follow, and methods for controlling them.

Epidemiologic studies of viruses borne by arthropods (which include crabs, insects, and the like) have led to the discovery since 1940 of over 200 new viruses among these lower animals, some of which cause disease in man. One is the California virus that causes encephalitis; others cause serious diseases in Africa, India, Southeast Asia, and the Mideast.

A notable discovery that polioviruses can propagate in tissue culture and cause visible destruction of cells has been the basis for major advances in understanding virus diseases in general. The discovery has facilitated the development of vaccines that have brought poliomyelitis and measles under control in all developed countries.

Many new respiratory viruses have been found and their patterns of infection and disease potential have been recognized. Some 70% of respiratory disease, nonbacterial in origin, is now known to be caused by viruses presently identified. Because influenza viruses can change their biochemical properties periodically, the World Health Organization has established a world-wide surveillance system to detect these antigenic character changes early. Appropriate vaccines can be developed to combat impending epidemics.

A virus has been found that can be transmitted venereally and may be related to cervical cancer. Another virus associated with a particular kind of abnormal growth and also with cancer of the nasal passages appears to be the long-sought cause of infectious mononucleosis.

Viral hepatitises, although not separable clinically, have been found epidemiologically to be of two kinds, incompletely separated in terms of the means by which they are transmitted. One type (serum hepatitis), so far as it can be distinguished, is transmitted largely through nonalimentary inoculations of blood, blood products (blood components), or materials contaminated with blood serum; but it is increasingly recognized that this type of hepatitis can be transmitted also by other means, presumably by fecal–oral routes. The other type (infectious hepatitis) is transmitted mainly through the fecal–oral route. The spread of serum hepatitis has been reduced through community health measures that include the screening of blood donors for transfusions and the use of disposable syringes and needles. The disease can be kept at subclinical levels by the use of gamma-globulin, a protective fraction of serum that contains immune bodies.

A recent advance has been the demonstration that patients with serum hepatitis develop a specific antigen (possibly the virus) in their serum, which also may be present in chronically infected carriers. The control of infectious diseases results from community-wide epidemiological and public health programs.

Additional knowledge and community health control measures have been developed relative to a type of pneumonitis (inflammation of the lungs) in man, caused by an agent harbored by nearly all wild and domestic birds. An important discovery was the agent (a rickettsia) causing Q fever, widely distributed in the United States and throughout the world, which comes not only from ticks but also from chronically infected livestock. During the breeding season, the livestock shed large amounts of the rickettsia and contaminate the environment. Stimulated by World War II, control methods have been developed for other rickettsial diseases, including vaccines against epidemic typhus fever, which caused widespread death after the first World War. The insecticide DDT, effective against the body louse that transmits typhus fever, has also helped to eradicate the disease.

Much of the advance in control of infectious and parasitic diseases has resulted from the greatly augmented armamentarium of drugs and biological agents (vaccines) that have become available particularly during the last 30 years. Because of these advances acute infectious diseases have declined in importance in developed countries. Epidemiologic investigations have now been directed increasingly to chronic progressive diseases that usually are considered to be noninfectious. Evidence is accumulating that some of these diseases (some types of cancer and several diseases of the central nervous system) may be caused by viruses. The role of infectious agents in chronic diseases is an area of research activity in which important advances may be expected in the future.

Other important advances in epidemiology and community health include evidence that "hardness" of water (due to calcium salts) is associated with a lower incidence of cardiac disease, somewhat as an optimum amount of fluoride is associated with healthy teeth. If so, it may be that in the future addition of calcium salts to soft water will help to decrease the prevalence of heart disease. A national Epidemic Intelligence Service has been developed to detect and avoid disasters exemplified by the thalidomide tragedy, the accidental production of paralytic disease by early polio vaccines, the leukemia-causing effect of excess ionizing radiation due to occupational or diagnostic exposure, the worldwide epidemic of retrolental fibroplasia (often leading to blindness) in premature infants given too high a con-

centration of oxygen, and poisoning resulting from consumption of grain treated with mercury.

Additional concerns of epidemiologic research, important to community health, are occupational diseases and hazards of an environmental character, such as air pollution, food additives, accidents, and cigarette smoking. Recognition of the important role played by cigarette smoking in lung cancer and coronary heart disease is a dramatic example of modern epidemiology.

MILITARY RESEARCH CONTRIBUTIONS

Many military research developments relate to community problems, air pollution and purification, sealed environments, hyperbaric oxygenation, radiation, human factor studies, medical instrumentation, and food technology. Studies initiated to solve military problems have provided tolerance limits for ozone, nitrogen dioxide, and carbon dioxide that have value in establishing community and industrial clean air standards. These studies have shown that pulmonary irritants interfere with the normal defense mechanisms of the human body against infectious agents by decreasing the mucus flow that removes microorganisms from the pulmonary air passages. Continuing research on the influence of ozone, nitrogen dioxide, and fluorine compounds, all common air pollutants, on upper respiratory infections is expected to throw light on the high death rate of old people during periods of acute air pollution. Carbon monoxide, a pollutant of air both indoors and outdoors, is being studied relative to the limits of human tolerance in exercise, and its potential role in cardiovascular complications.

Studies on space and aircraft materials, that both do and do not contribute to environmental contamination, have provided data useful for the selection of nontoxic and noninflammable materials, not only for aircraft but also for hospitals and other structures. Research on the performance of humans in sealed environments, as in space and on the ocean floor, has led to techniques for the purification of air in small enclosures applicable to hospital patients requiring ultraclean air. A new laser technique for detecting trace metals has been developed and used in a cooperative military–civilian study of industrial workers exposed to beryllium. There is a growing awareness that trace constituents in expired air should constitute an excellent early indicator of disease prior to clinical symptoms, an approach that should lead to useful developments in the future. Reliable and sensitive detectors have been developed that can be applied in monitoring the quality of the environment. Toxicological and pharmacological information on new fire extinguishing agents applicable to community problems has been obtained.

Military research on the effects of mechanical forces on man have yielded principles applicable to civilian problems. Research on crash injury and protection has led to joint military–civilian studies that have improved protective devices used in the communities. Criteria relative to noise exposure and protection, developed by the military, have been the major basis for international standards of industrial and residential noise limits. Military and civilian personnel have worked jointly to derive acceptable operational procedures by which to minimize the effects of sonic boom. Principles and methods for evaluating hearing devices have been applied in the civilian community.

Studies of the effects on personnel of vehicles such as tanks, helicopters, and military aircraft helped to provide a basis for standards of industrial truck, railway, and other transportation systems, and various military methods of cushioning and of body protection have influenced the form of civilian counterparts. Other research has contributed to patient care techniques; for example, certain patients with cardiovascular disease sometimes are not able to perform normal exercises or even remain erect because their circulating blood is unable to return from the extremities to the heart against the pull of gravity. A similar situation occurs with aviators exposed to forces several times greater than gravity and so they wear an antigravity suit that applies counterpressure to the body surface to prevent pooling of blood in the extremities. This has led to the development of elastic garments for use by cardiovascular and other patients. Human factor studies aimed at fitting human beings most effectively and most protectively into small enclosed environments, such as cockpits, have resulted in consistent advancements in the state of this art.

These studies have provided new and more comprehensive data on body size and the mechanical aspects of body motion, have provided new standards, and have advanced measurement techniques. The developments have had widespread application in such health-related activities as the design of automobiles and aircraft, special clothing, protective equipment, artificial limbs, and patient care equipment for use in major medical centers. Human engineering studies had results with a number of applications, as have studies on human visual function and on human task performance as affected by multiple stresses such as fatigue, lack of sleep, noise, and other disturbances. All these developments have led to medical and health standards that are widely applied in preventive medicine, public health, and epidemiology.

DRUGS

In 1940, the physician's "black bag" was relatively small. If it did bulge, it was apt to be filled with many carryovers from the "weed and seed" days of medicine. Camphor was still used as a circulatory and respiratory stimulant, Coramine and lobeline were popular respiratory stimulants, and bromides were common sedatives. Thiocyanates were used to treat hypertension. Bile salts were widely used for a variety of vague indications of ailment. Colloidal silver preparations were used in the treatment of gonorrhea, upper respiratory infections, and bacillary dysentery. Gold was the treatment of choice for systemic lupus erythematosus, a disease involving eruptions and scarring of the skin, with no more than a third of treated patients reported cured. Mental hospitals were packed, and there were virtually no drugs for the treatment of mental illness. Psychotic patients were often committed to obsolete back wards of institutions and destined to remain there for their lifetime.

Physical restraint, hydrotherapy, electric shock (or insulin), and barbiturate sedation were virtually the only treatments—along with psychotherapy. With the discovery of the major and minor tranquilizers and the antidepressants, some 60,000 fewer hospital beds are needed today for patients with mental illness.

There have been more significant drug discoveries and developments during the past 30 years than in the total previous history of man. In 1940 there were available no antibiotics; no antihypertensives for the treatment of high blood pressure and hypertension; no oral diuretics; no contraceptive pills; no antihistamines; no oral hypoglycemics for use in conditions of high blood sugar such as occurs in diabetes; no significant psychopharmacologic drugs for use in mental illness; no antiemetics to combat seasickness, airsickness, and vomiting; no corticosteroids for use in arthritis and other disorders; no radioisotopes for use in diagnosis or treatment; few or no chemotherapeutic agents for use against viral diseases or cancer. There were no drugs to reduce hyperlipemia (excess lipids or fatlike substances in the blood); no vaccines to prevent polio, German measles, measles or mumps. There was no L-dopa for Parkinson's and related diseases (involving paralysis and shaking); no isoproterenol for cardiogenic shock (such as is associated with severe heart attack); no nonnarcotic pain relievers; no potassium-retaining diuretics; and no nerve beta-blocking agents. All of these, and more, are available to physicians today.

World War II stimulated at least three major drug research efforts that gave a massive impetus to drug discovery and development: an extensive antimalarial chemical and pharmacological search, necessitated by restriction of quinine importation during the war; a crash program for the commercial production of penicillin (mentioned above); and the synthesis of cortisone in 1944 (first used clinically in 1948). The latter accomplishment was spurred on by a false rumor that German pilots were using a new antishock hormone that not only enabled them to fly at higher altitudes but also helped their wounds to heal more rapidly.

When cortisone was introduced commercially in 1949 it cost $200.00 per gram; today it costs about $1.25 per gram.

In the last decade the rate of new drug development has seriously declined, perhaps due in part to legal requirements of proof of efficacy and of safety for all new drugs. In 1960, 306 new drug products appeared in the United States but the number has steadily decreased until in 1969 only 62 appeared. To "prove" each new drug efficacious and safe requires an estimated 7 years in time and costs over 5 million dollars. It is highly unlikely that there will ever be a drug entirely free from side effects and completely devoid of some risk.

IMMUNOLOGY

Imagine a geographical territory that includes villages inhabited by natives who look with such disfavor on strange persons or animals that all such that enter the territory will be destroyed. A strange animal appears and the natives, if they do not already have at hand effective weapons with which to slay the creature, quickly make them and with these new weapons attack and slay the invader. Something of this sort is characteristic of an immune reaction, which is a destructive reaction by the native tissues of an organism, utilizing their manufactured weapons (antibodies) against the infectious invader and his provocative antigens. Most or all of the infectious diseases and many if not all of the parasitic diseases to which reference has been made thus far involve these chemical immune reactions in the bodies of infected persons. But the immune reaction is actually a complicated process, not yet completely understood. During the past 30 years, however, great advances have been made in the understanding of what does take place in the bodies of patients with diseases involving immunologic disturbances. That knowledge has led to the development of improved methods of prevention and treatment, as indicated by the control of infectious diseases by new and improved vaccines already discussed above. The revolutionary developments in immunologic techniques have also yielded diagnostic procedures useful in other diseases.

Among the important advances that have been made in understanding the events that take place in immunologic reactions are these: more is now known about the synthesis, structure, and biological action of the various antibodies, about the particular cells that mediate immunity, and about the role of the lymph glands and the thymus in immune reactions. A greater insight into immunologic tolerance has provided a basis for manipulation and control of the immune process, a matter of the greatest importance for surgical transplantations. The frequency and disease potential of the auto-immune process have been determined, a process in which an organism directs the immune reaction against its own tissues. How the immune processes may actually cause disease has become much better known; the immune reaction is now known to involve a complex sequence of molecular and cellular events that produce injurious substances to alter function and cause tissue damage in one or more areas of the body. Even more important, it is known now that the immune process can be blocked, potentially at least, at each step in the sequence and that immunologic disease is, at least to that extent, susceptible to treatment. With this increased knowledge more effective vaccines have been developed for the control of a number of infectious diseases. Several have been greatly lessened in severity and some virtually eradicated (Fig. 2 and Table 1). In addition to the blockage method, control of immunologic disease has been accomplished by the administration of antibodies from one person to another person who is affected by a disease (passive immunization), an outstanding example of which is the administration of anti-Rh antibodies to Rh-negative mothers to prevent Rh disease. Passive immunization has also been used to prevent or modify certain virus infections, including hepatitis, measles, mumps, and rabies.

Techniques useful in immunologic diseases have increased in the past 20 years: methods of augmenting or suppressing immune reactions, of substituting for missing immunologic components, and of blocking the action of biochemical agents or cellular components that cause the injury. The most used method, suppression of the immune process or of its effects by some chemical agent or hormone or appropriate antibody, is used when an individual's own immune system is causing injury to his own body; also it is this method (immunosuppression) that has made transplantation of organs possible. Augmentation of the immune reaction is employed when necessary to stimulate protective antibody formation, including the formation of interferon to protect against a variety of viral infections. Substitution of missing immunologic components makes the difference between normal life and death from repeated infections for individuals with either inherited or acquired immunological deficiencies. A new form of substitution, called "cellular engineering," is now being developed involving the transplantation of certain glandular cells and bone marrow, and even the supplying of chemical factors to initiate missing cellular processes.

The new immunologic technology, which has made such significant strides in the past 30 years, is now widely used and provides the basis for a great part of diagnostic laboratory medicine today, including methods such as blood typing for transfusions, tissue typing for transplantation, genealogic identification of infectious agents, and identification of antigens in a variety of cancers. One of the most dramatic as well as important achievements of clinical immunology is in the area of organ transplantation.

TRANSPLANTATION

Prior to World War II the problems of organ transplantation were not even vaguely anticipated. Between 1940 and 1945 the foundation for subsequent work in this field was laid by the demonstration that graft rejection was due to an immunologic reaction of the host to the foreign tissue of the graft. The key observation in support of this concept was that a second skin graft from the same animal that donated the first was destroyed (rejected) more rapidly than was the first graft. This led to the idea that, as a result of the first graft, the host must have developed a kind of immunity against the foreign tissues of the graft. If this were so, why not prevent this immunity or at least weaken it?

Each new kind of treatment that could be shown to weaken immunologic defenses was tried out in animal experimentation, including treatment with agents such as total-body irradiation, certain steroids, and certain antimetabolites. Transplantation was not successful when only one of these agents was used alone. Overtreatment killed the host and undertreatment did not prevent rejection of the graft. Only when such suppressive agents were used in combination did organ transplantation become a practical reality.

Before 1962 there were only two really long-term survivors of renal transplantation. These had received kidneys from fraternal twins and also had been given whole-body irradiation. The modern era of clinical organ transplantation dates from 1962 when extensive trials were begun with various combinations of agents. With this new approach the air of pessimism changed to one of optimism. It was quickly demonstrated that rejection of a graft could be reversed easily with proper manipulation of these agents, particularly by adjusting the steroid doses. Moreover, it was learned that when the first wave of graft rejection was prevented or reversed, the degree of subsequent immunosuppression could often be reduced, so that patients could be maintained on smaller doses of the suppressive drugs. It seemed that some specificity of effect had been achieved even though the agents being used were overall immune suppressants. It is anticipated that the consequent altered relationship of host to graft will become better understood as a result of future research and that these techniques will be further improved. Whatever the explanation may be, one thing has been thoroughly established; a graft from one person to another can come to be accepted in its new and previously hostile environment within the body of the host.

In 1962, and for several years thereafter, a number
of groups entered the field of renal transplantation.
This treatment is now provided by surgeons at the majority of American medical schools. The long-term
benefit of renal transplantation has been demonstrated,
especially if the transplanted kidneys are donated by
blood relatives. About fifty patients treated in 1962
and 1963 are still alive and in good health, 8–10 years
after transplantation. Several thousand subsequent
recipients of kidneys are now surviving and live essentially normal lives.

Research on renal transplantation continues in an
effort to improve methods of immunosuppression and
of donor selection. Since 1966 several new potent im-
munosuppressants have been discovered, and evaluation of these is still underway. By the use of tissue typing
and cross matching of immune cells, physicians are
seeking organs that will provoke a weak immunologic
reaction. It is anticipated that progress will be made
along both these lines.

With these successes in renal transplantation it became increasingly obvious that other kinds of organs
could be transplanted successfully if the principles
learned from kidney transplantation were applied. The
liver was the next organ to be transplanted successfully,
using a combination of immune suppressants. With
the same immunosuppressive techniques came the first
chronic survivals after human heart and lung transplantations. Although the successes to date have been
relatively few with transplantation of organs other than
kidneys, the feasibility phase is now complete. It is to
be expected that, in the future, progress will be made
in increasing the survival time of the patient and in the
transplantation of other organs such as the pancreas
and the bone marrow, initial successes with which have
already been reported.

Organ transplantation has been successful because
and only because of the knowledge made available by
basic research on immunologic mechanisms, on techniques for their control, on tissue and organ preservation, on the physiology of denervated complex organs,
and by new surgical techniques derived from animal

experimentation. This is the real achievement of organ transplantation, dramatized by the spectacular surgical act of transplanting the physical organ itself.

Although so heavily involved with things immunologic, organ transplantation is a surgical operation and, like other special surgical areas, is much concerned with anesthesiology, which within the past 30–40 years has become a recognized medical specialty.

ANESTHESIOLOGY

For almost a hundred years after the introduction of surgical anesthesia its use was unscientific and often left to inexperienced medical assistants. In the early 1930's a group of American physicians, located in various parts of the country, become interested in anesthesia as an underdeveloped area of medicine. They formed an Anesthesia Travel Club. These were the pioneers in the now recognized field of Anesthesiology, for which they established training programs.

About that same time a new anesthetic, the gas cyclopropane, was introduced. Its administration required greater skill and a more thorough knowledge than did either chloroform or ether. It was so different that the interest of physicians was aroused. When it was noted that the pulse became irregular and respiration depressed during cyclopropane anesthesia, physicians sought the aid of pharmacologists, only to find out how little was known of the pharmacology of anesthetic drugs. Even the basic pharmacology of ether was unknown, although ether had been used since 1846. Simultaneous studies of cyclopropane and ether were undertaken beginning the early stages of clinical pharmacology.

Cyclopropane could not be administered by open methods, and this led to the acceptance of the closed system of anesthesia that had been used only occasionally with ether. The closed system led to the development of flowmeters and carbon dioxide filters, and to research on absorbents such as soda lime and barium hydroxide mixtures. Research on the effects of anesthetics included their uptake, distribution, and fate

in the body, their action in causing irregular heart
rhythm, and the effect of combining the anesthetics
with various adjunctive drugs. The first basal narcotic
to receive extensive study was tribromoethanol solution,
administered rectally. Later the barbiturates were
classified and their effects were investigated in great
detail. Methods of regional anesthesia and of prolong-
ing anesthetic blocking action also by the use of drugs,
causing constriction of blood vessels, have been improved or perfected during the past 20 years. The use
of these blocking agents has been made safe as a result
of studies on the physiology of the nervous system.

The research anesthesiologists, pharmacologists, clinicians, and others, have investigated various narcotics
and semisynthetic opiates, introduced the use of anti-
narcotics to reverse narcotic depression, shown the advantages and disadvantages of certain nerve stimulants
used to overcome depression due to drug-induced coma,
and obtained new information about the effects of
drugs on the lungs. Also they developed improved
methods of artificial respiration and improved mechanical ventilators for long-term artificial ventilation of
the lungs.

The anesthesiologists have made important contributions to cardiovascular disease and obstetrics and
have introduced adjuncts to anesthesia such as muscle
relaxants, thereby permitting lighter anesthesia while
affording optimum operating conditions for the surgeon.
The muscle relaxants have been studied from the standpoint of their classification, pharmacological action,
detoxification and elimination, and the effects of electrolytic and genetic factors on their action in the body.
The effects of anesthetics on blood flow through the
lungs, brain, and heart, on the gastrointestinal tract,
liver and renal function, and on the obstetric patient
and the newborn infant have been investigated. More
is now known about the passage of drugs across the
placenta, their distribution in both mother and fetus,
and their effect on labor and uterine motility.

During the past 30 years anesthesiologists have introduced monitoring techniques to improve patient care

and safety, such as measurement of pulmonary ventilation, blood gases, and arterial and venous pressure, and the recording of brain waves and electrical changes in muscle contraction. They have worked with engineers and fire control officers to prevent explosions and fires in surgical suites; with surgeons to improve techniques for monitoring blood flow during anesthesia and for determining the effects of drugs; with neuropharmacologists on the effects of psychosedatives on the brain and nervous system; and with internists and surgeons on the interaction of anesthetic drugs with drugs given the patient prior to surgery. The anesthesiologists have contributed importantly to the development of postoperative recovery rooms and intensive care units.

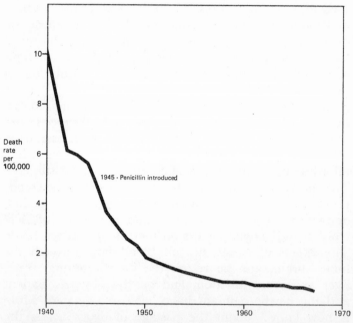

FIG. 3. Age adjusted death rate per 100,000 population from appendicitis since 1940. *Source:* Office of Health Statistics Analysis, National Center for Health Statistics, Health Services and Mental Health Administration, Public Health Service, Department of Health, Education, and Welfare (1971).

When ether anesthesia, America's early and great contribution to the development of surgery, banished pain in compulsive surgery, it did not expand elective surgery. This advancement came only with the later acceptance by surgeons of the goal of germ destruction (antisepsis). The modern era of hospital and operating room antisepsis has come about during the past 30 years.

Because of the extensive splintering of surgery into specialized areas, such as cardiac surgery, general surgery is now a mere fragment of its former self. Many operations classified as general surgery are no longer necessary because of antibiotics and improved medical care. Between 1934 and 1967 deaths from appendicitis fell from 16.4 to 0.08/100,000, a reduction of 95% (Fig. 3). In this country the sharp decline began about 1940–1942, prior to penicillin, which first became available in small amounts for civilian use in mid-1944. This decrease in mortality was due essentially to three factors: the avoidance of purging in all acute abdominal lesions, a better understanding of the nature of appendicitis that became recognized as obstructive and not inflammatory in origin, and early recognition of the disorders together with improved surgery and anesthesia. In addition, morbidity and mortality from surgical removal of the appendix have been decreased by the use of antibiotics.

Since 1940 there has been a decrease from the formidable figure of 60% to 2–5% in mortality caused by obstruction of the viable and unperforated intestine. This reduction was associated with improvement in techniques of intestinal decompression, either by surgical operation or by tubes passed through mouth or nose into the stomach and through which suction was applied. Similarly, mortality from congenital or pathological closure of the intestine has been greatly reduced. The chief persisting cause of mortality from acute intestinal obstruction today is loss of viability due to blood vessel occlusion brought about by such agencies as adhesive bands, twisting of the intestine, and strangulated hernia.

CARDIOVASCULAR SURGERY

While general surgery was undergoing its happy decrease, cardiac surgery grew rapidly. As a therapeutic activity and a surgical subspecialty, it has found fruition since World War II. Many of the surgical techniques employed in cardiovascular surgery today, even techniques of cardiac transplantation, were developed during the first decades of this century but little or no clinical application was possible at that time due to lack of the current methods of anesthesiology, such as the passage of an anesthetic gas into the lungs by way of a tube inserted in the trachea or windpipe. Nor were present day techniques of blood transfusion available. A successful operation on the mitral valve of the heart was performed as early as the 1920's. This is an operation in which the surgeon cuts through the valve to enlarge its opening when it has become so small as to interfere with adequate circulation of the blood. But at that time the medical climate was not yet ready for such bold departures.

World War II helped pave the way for present day cardiovascular surgery. A better understanding of cardiovascular and cardiopulmonary physiology resulted from the management of large numbers of battlefield wounded, with injuries of the chest, heart, and blood vessels. Blood-typing and blood-storage techniques were refined and further developed in answer to the needs of surgeons caring for these war wounded. In addition, the numerous research programs in medical centers began to bear fruit following World War II leading to sudden and explosive developments in our understanding of cardiac surgery.

Foremost among the early developments was a clinically successful operation performed in 1944 as a surgical treatment for a congenital defect of the heart called the "tetralogy of Fallot." In this condition, not enough blood circulates through the lungs and so not enough oxygen is taken to the tissues of the body. A

child born with this defect, a so-called "blue baby,"
did not live very long and was not well while he lived.
What the surgeons did was to sew together one of the
chest arteries (the subclavian, under the collar bone)
to the pulmonary artery that carries blood to the lungs,
thereby enabling enough blood to pick up oxygen from
the lungs to supply the body tissues adequately (Fig. 4).
The life of these patients was greatly prolonged. That
same year (1944) brought a demonstration that a
congenitally narrowed aorta, the main and largest
artery carrying blood from the heart, could be repaired
successfully by surgery. A surgeon cut out the narrow
section of the aorta and then sewed together the two
cut ends of the severed vessel. Four years later, in 1948,
another surgeon showed that instead of sewing together

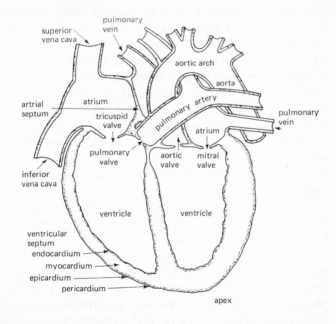

RIGHT SIDE OF HEART   LEFT SIDE OF HEART

FIG. 4. The human heart. *Source: A Handbook of Heart Terms.*
National Institutes of Health, Public Health Service, Department
of Health, Education, and Welfare (1968).

the two cut ends of the severed aorta, the repair could be made by using a segment of aorta from another person; that is, he used an aortic graft. This development altered the course of surgery for numerous vascular diseases.

The year 1948 also brought advances in control of both acquired and congenital cardiovascular diseases. It was shown that something could be done about a rheumatic heart, a heart with a defect caused by rheumatic fever resulting in a narrowing of one of the main valves of the heart (the mitral). Because of the immune process that occurs during rheumatic fever, which is caused by infectious bacteria (streptococcus), the flaps of the mitral valve become inflamed, become enlarged, stick together, and develop scar tissue, causing the valve opening to become too small. Strands of tendinous tissue are normally attached to these valve flaps, somewhat like ropes attached to the edges or corners of sails on a boat, and help to control their movement. In the case of a rheumatically narrowed valve, cutting or stretching of these fused tendons or commissures apart, a process called commissurotomy, helps the valve to open more widely and to let the blood through more effectively. This accomplishment stimulated interest in mitral commissurotomy as a surgical treatment for mitral valves made defective by acquired disease. Then a technique was developed by which a surgeon could sew up (suture) congenital holes in the thin wall separating the two smaller chambers of the heart (the two auricles).

The pace of advancements in cardiovascular surgery became even more rapid in the first half of the 1950's. Earlier, in 1929, an investigator had demonstrated (on himself) that it was possible to insert a small flexible tube (a catheter) into a vein and pass it up through the veins into the chambers on the right side of the heart, a procedure that became known as cardiac catheterization. Techniques of cardiac catheterization were refined and applied clinically, an achievement for which the Nobel Prize in Medicine was awarded in 1957. Such techniques allowed extension of clinical diagnostic

methods and assured the surgeon of the anatomy and
physiology he would encounter at the time of operation.
Another achievement, the development of techniques
of hypothermia, enabled the cardiac surgeon to operate
on a heart without the interfering complication of heart
beats and flowing blood. Hypothermia safely lowers
body temperature, and is achieved by packing the body
of a patient in ice. Normally it is injurious or fatal to
lower body temperature excessively but by first giving
a patient a sedative his body temperature can be lowered and maintained below 90 F for as long as several
days. While thus artificially cooled, the patient's heart
and blood circulation can be stopped, for 10 min, allowing the surgeon to deal with a still, dry, and open
heart. Cooling in this controlled way is not in itself
injurious.

Repair of a congenital defect in the wall separating
the two auricles of the heart was shown in 1953 to be
surgically possible, the operation being done under
direct vision while the blood was made to bypass the
heart by circulating through a perfusion-pump apparatus outside the body and back into the body again.
This technique of extracorporeal circulation with an
oxygenator for aerating the blood was the cumulation
of more than 2 decades of work on the pump oxygenator. This technique opened the way for modern-day
heart surgery.

Atherosclerosis, a cardiovascular ailment in which
fatty or lipid-containing material is deposited in the
inner part of the walls of blood vessels, and a major
cause of death on the North American continent (Fig.
5), also came under surgical attack in the early 1950's.
In the 1920's a technique had been developed whereby
X-ray pictures of blood vessels and other cavities could
be taken after the cavities had been injected with
material opaque to X rays, making them visible in
photographs; but this technique had remained primarily within the province of the neurosurgeon interested in the blood vessels of the brain, and the urologist
interested in the kidneys and bladder. The radiographic
method was adapted in the 1950's by cardiovascular

surgeons to the mapping of the aorta and other arteries, i.e., the arterial tree. In 1951 a surgeon cut out an atherosclerotic section of aorta in the abdomen that had become abnormally dilated and replaced it with a section of normal aorta from another person, a grafting technique of revascularization similar to that employed in 1948 for congenital narrowing of the aorta. This technique was soon duplicated by others who extended it to dilations of the thoracic aorta and almost every portion of the arterial tree. Methods were developed for using grafts of parts of normal blood vessels, taken

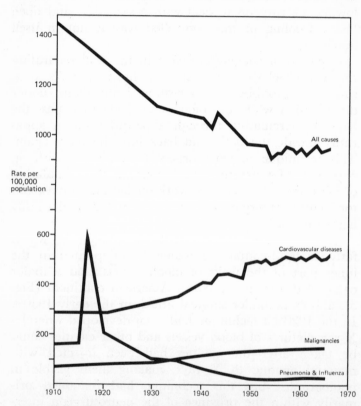

FIG. 5. Leading causes of death in the United States since 1910. Rate per 100,000 population. *Source: Health, Education, and Welfare Trends*, 1966–67 edition. Department of Health, Education, and Welfare (1968).

from other persons, for revascularization in patients
with blood vessels obstructed with atherosclerotic material. Recognition of certain features of atherosclerotic
involvement was significant in developing vascular
surgery. Primary is the concept that although atherosclerosis is a generalized systemic disease, involving
most of the circulatory system, in its natural course it
involves some areas of the body more than others.
Atherosclerotic lesions tend to be localized even though
multiple locations may be involved. A localized atherosclerotic lesion, whether it stops up the vessel (occlusion)
or is associated with abnormal dilation and weakening
of the vessel wall (aneurysm), usually is located above
or below the diseased section. The diseased section can
be excised and replaced surgically with a resulting re-
establishment of relatively normal arterial blood flow.
This surgical procedure is usually successful.

Advances in vascular surgery were further facilitated
by the development of satisfactory grafts made of synthetic material for arterial replacement. This eliminated
the need for grafts obtained from other people, grafts
that were in short supply and involved significant long-
term complications. Synthetic materials, Dacron and
Teflon, were used to repair the heart itself, which now
could be kept open for prolonged periods of time
through the use of the pump oxygenator. Patches of
the synthetic were utilized for closing holes within the
heart, and to enlarge certain other areas of the heart,
as in severe forms of tetralogy of Fallot mentioned
above. Toward the end of the 1950's leaflets of these
materials were used to replace one, two, or all three of
the flaps of the aortic valve (at the junction of the aorta
and the heart) when they became irreparably diseased.

The past decade has been particularly exciting for
cardiovascular surgery. The majority of congenital
cardiac lesions now can be corrected. Even transposi-
tion of the great vessels, a common congenital defect
which usually kills in infancy, has been brought within
the category of operable diseases by the development of
special surgical techniques. Advances in the surgical
correction of acquired cardiovascular disease have been

even more phenomenal. It became apparent that long-term results of cardiac valve replacement with leaflets of synthetic fabric was frequently unsatisfactory. This was because eventually fibrous tissue would grow on the fabric clogging the valve opening, or the fabric would become weakened and tear. The use of synthetic fabric for artificial valve flaps was replaced by a different kind of artificial valve arrangement that consisted of a ball within a valvular cage. As early as 1951 this principle was applied by a surgeon who used an artificial caged-ball valve in the descending aorta (in the chest) in selected patients with leaky aortic valves. In 1960 an artificial caged-ball valve was used to replace a diseased aortic valve itself and also a diseased mitral valve in a human heart. Thus was born, for all practical purposes, the field of clinical cardiac valve replacement surgery. Rapid and progressive modifications have significantly improved surgery and its effect on the blood and on the character of blood flow through the valves, and their complications have been reduced to an acceptable minimum. Heart valve replacement can be recommended today whenever a patient begins to show symptoms of impaired valve function.

The 1960's also saw the advent of implantable cardiac pacemakers for treatment of various forms of deficiencies in the electrical system of the heart. This is the system by which nervelike impulses are conducted from the normal pacemaker to the heart muscle during regulation of the normal rate of heart beats. Early models of the artificial pacemaker gave trouble because of breakage of leads, failure of component parts, and premature battery failure. Rapid advances in bioelectronics now have reduced these problems to an acceptable minimum. Two patients in France recently have received implanted cardiac pacemakers powered by radioactive plutonium.

The most spectacular of the developments in cardiovascular surgery within recent years has been transplantation of the entire heart, first achieved in December 1968 using techniques developed in the early 1960's. An outbreak of similar procedures followed at medical centers throughout the world. These transplantations were done with the hope that the immunosuppressive techniques developed in connection with renal transplantations would be satisfactory also for cardiac transplantation. It was hoped also to learn more about the heart's normal and abnormal physiology, the demands placed on the diseased heart, the demands to be placed on the transplanted heart, and also the demands that would have to be placed on mechanical devices that might come to be used for total heart replacement.

As one reviews significant developments in cardiovascular surgery since the end of World War II, it is well to point out a number of previously discovered techniques which may be applicable to untold hundreds of thousands of patients suffering and dying from atherosclerosis of the coronary vessels, the vessels within the walls of the heart itself. X-Ray methods of visualizing the coronary arteries now provide an accurate map of the coronary circulation and its collateral branches. Use of the pump oxygenator allows extensive manipulation of the heart and the coronary arterial system while maintaining the remainder of the body in good condition. The principle of bypassing localized diseased sections of vessels provides a means of correcting the bad effects of atherosclerotic lesions in the coronary vessels in a large number of patients. These techniques can be used until such time as intensive investigations currently under way provide an understanding of the cause of the disease. Until then, or until we can satisfactorily replace the entire heart, vascular graft bypasses from the ascending aorta to the distal coronary arterial tree promise to be a most useful cardiovascular surgical operation.

CARDIOVASCULAR DISEASE

Since 1940 there has been an increasing interest in diseases of the heart muscle, called cardiomyopathies. The discovery that many toxic substances and various disease states could produce abnormalities in heart muscle itself initiated investigations that still continue.

A disease of heart muscle that obstructed the flow of blood from the heart into the aorta, the largest vessel carrying blood from the heart to the body, was described for the first time in the late 1950's. Of particular interest is evidence that alcohol alone can cause heart disease. Prolonged bed rest was found in the late 1940's to be beneficial to patients with heart muscle disease. The influence of climate was recognized and adequate heating and air-conditioning was used to improve patient care. Much new knowledge was gained during the 1960's on the finer structure of heart muscle, its metabolism, and its mechanism of contraction, in relation to problems of heart muscle disease.

Medical research from the military establishment has provided results of significance in several areas. Researchers have studied the effects of acute and chronic exercise, physical inactivity from prolonged bed rest, and long-term repeated exposure to peak-capacity exercise, with other stresses (as heat, cold, altitude) superimposed. The investigations apply to civilian problems such as bed rest in medical treatment, physical conditioning as a preventive medical measure, and the rehabilitation of patients. Evaluation of serial electrocardiograms from hundreds of thousands of normal men has greatly increased the understanding of what is and what is not normal, and of the significance of changes in the same individual.

In peripheral vascular diseases, the development of plethysmographic techniques (methods for measuring changes in the volume of an organ or, for example, an arm or a finger) have improved the diagnosis and understanding of normal and abnormal vascular physiology. Technical improvements in the late 1960's provided methods for the continuous measurements of body temperature, blood pressure, and electrical activity of the heart and the measurement of peripheral blood flow by the use of ultrasonic (ultra sound) vibrations.

New resuscitation techniques have saved many lives. These techniques include massage of the heart through the closed chest to maintain circulation of the blood, direct massage of the heart after the chest has been cut

open, mouth-to-mouth breathing that is more effective
than the traditional artificial respiration procedures,
and electric defibrillation of the heart, first performed
in 1947. When the heart fibrillates, its muscle fibers
undergo extremely rapid and disorganized twitching
but the heart as a whole does not function. Fibrillation
of the ventricles of the heart usually is fatal unless the
heart is quickly defibrillated. A modification of this
defibrillation technique called electric cardioversion, for
restoration of the normal heart rhythm, except in case
of ventricular fibrillation, was developed in the 1960's.
In the late 1950's and 1960's devices were developed
for "pacing" the heart in cases of complete heart block,
a condition in which the nervelike impulses causing the
heart muscle to contract do not get from the auricles to
the ventricles. The first permanent implantable artificial pacemakers were put into patients in 1960.

As mentioned elsewhere, although very recently there
has been a disturbing increase in syphilitic infections,
after the advent of the antibiotics syphilitic heart disease and syphilitic widening and weakening of the wall
of the aorta virtually disappeared. Rheumatic heart
disease became much less common, and the once 100%
fatal subacute bacterial endocarditis became almost
100% curable. New and more potent diuretic drugs,
developed in the 1950's and 1960's, have been helpful
in cases of severe congestive heart failure, renal edema,
and arterial hypertension. Since 1940 a number of
anticoagulant drugs have been introduced, including
one that can be taken orally and its dosage controlled
by a simple laboratory test. Many new drugs for the
control of heart rhythm were developed in the 1960's.
Although definitive treatment for atherosclerotic cardiovascular disease has not yet been found, certain
"risk" factors have become recognized, such as elevated
blood pressure, lipids or fats in the blood, obesity, lack
of regular exercise, cigarette smoking, and family history of the disease.

Among the important techniques that have been
developed are right heart catheterization (mentioned
above) that permits measurements of the outflow of

blood from the heart, direct puncture of the left side of the heart to get accurate X-ray pictures, and similar injection techniques for use with the blood vessels. Other advances in X-ray picture techniques provide X-ray moving pictures and rapid successions of still pictures that permit a better study of occlusive vascular diseases, congenital defects, and other disorders of the vessels, including those of the cornorary arteries. Devices have been developed for recording and analyzing heart sounds and movements of the chest and also movements of the whole body caused by the beating heart. Electrocardiography had been well developed by 1940. Significant advances have been made since then, including the highly sophisticated techniques of vector cardiography some of which are three dimensional, a development that may have more clinical use in the future. Radiotelemetry, the recording of electrocardiographic and other signals from a distance even as far as from a man on the moon, has recently been advanced and in the future will have greater clinical application.

When the blood pressure rises to a sufficiently high level and remains high, a person is affected by the cardiovascular disease hypertension, of which there are several kinds; one is associated with kidney disorder and is known as renal hypertension. As early as 1934 it had been demonstrated experimentally that constriction of the renal arteries can produce hypertension in animals. Subsequently, some patients with hypertension were found to be suffering from a narrowing of their renal arteries. This discovery has led to a means of curing some of these patients and also has provided a better understanding of the disease. Extensive investigations during the 1940's and 1950's disclosed the role that renin, an enzyme from kidneys deprived of part of their blood supply, and angiotensin, a substance formed when renin from the kidney acts on a particular protein of the blood plasma, play in kidney function, in electrolyte (salt) balance, and in the production of hypertension. Angiotensin is a hypertensive agent causing blood pressure to rise. The roles of other substances having vascular effects are also being studied relative to the control of blood pressure and circulation in health and disease. For example, aldosterone, a hormone from the adrenal gland, was found in the 1950's to be important in the development of hypertension in some people, and also in the development of a newly recognized disease that involves cardiovascular disturbances.

High blood pressure, well known to shorten the life of man, had no effective drug treatment until 1955. Then certain drugs were found which reduce arterial blood pressure; these included Rauwolfia alkaloids, long known to the natives of India. During the next 15 years many other drugs capable of lowering blood pressure were introduced, including some potent antihypertensive drugs that can be taken by mouth. Since these drugs were first supplied commercially the mortality from hypertension has been reduced by at least 40 % and the morbidity even more (Fig. 6). Considering the hundreds of thousands of people who suffer from high blood pressure, the development of these drugs is exceeded in importance only by that of the antibiotics.

Hypertension, as well as some other disease states, may involve edema, an abnormal accumulation of water in the tissues. Thirty years ago a physician began the practice of medicine with a paucity of aid to approach his biggest and most serious medical problem: hypertension and the management of renal edema. To be sure, the involvement of the kidneys and the nervous system in the development of hypertension had been demonstrated in laboratory animals. Mild hypertension, if treated at all, called for the prescription of phenobarbital and patients with severe or malignant hypertension were sometimes exposed to the debilitating effects of the surgical excision of certain groups of cells of the nervous system. This left the large proportion of hypertensive patients without effective or acceptable treatment. Treatment was attempted by rice diet and by the restriction of salt in the diet, and there were the toxic thiocyanate drugs for those whose illness warranted the risk; but acceptable treatment for hypertension lagged behind that for infectious diseases by a decade or more.

With one exception, the same was true for the management of edema. Thirty years ago it could be said that, except for the organic mercurial drugs, the diuretics (drugs causing increased urinary excretion) were least effective when needed most. The physiologist had known long before this about the more general relationship between salt and the reabsorption of water back into the blood stream from the kidney, but potassium management by the kidney awaited the late 1940's for an explanation. In spite of the great attention to salt and water balance, to urine volume, and to the use of organic mercurial drugs as diuretics, the situa-

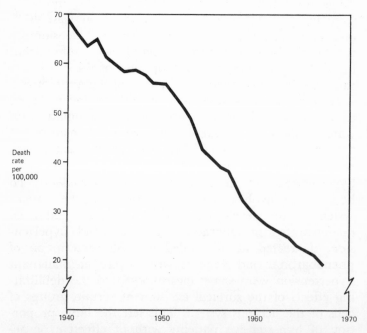

FIG. 6. Age adjusted death rate per 100,000 population from hypertensive diseases since 1940. *Source:* Office of Health Statistics Analysis, National Center for Health Statistics, Health Services and Mental Health Administration, Public Health Service, Department of Health, Education, and Welfare (1971).

tion was most frequently unsatisfactory for both the physician and the patient.

The turning point in the treatment of hypertensive disease was the development of chemical agents for blocking the activity of groups of cells (ganglia) of the nervous system. Pharmacologists succeeded in developing some ammonium compounds that could be used as ganglionic blocking agents in the treatment of hypertension. Skill in the use of these drugs was less well developed in the United States than in several other countries. The compounds were not well absorbed into the body and produced undesirable side effects. The best that could be said for these drugs was that they would lower blood pressure and that they did represent a turning point in the search for effective antihypertensive agents.

In the latter part of the 1940's an ancient recipe for combating high blood pressure, mental instability, and other less credible indications of ailment, came from India, bringing into Western medicine the Rauwolfia alkaloids derived from the roots of *Rauwolfia* plants. The drugs, particularly one called reserpine, displaced phenobarbital that was being used in the treatment of hypertension.

After a decade or so reserpine itself gave way to newly developed synthetic tranquilizers, particularly the phenothiazines. Reserpine remains an important element in the therapy of hypertension primarily because it has been an effective drug, but newer treatments, more effective and more acceptable to physician and patient alike, have limited its use. Reserpine contributed to a better understanding of antihypertensive therapy when it was found to release certain chemicals from nerve endings, knowledge important for an understanding of hypertension caused by nerve factors.

Antihypertensive treatment was transformed by the introduction of another group of drugs, the thiazides, unquestionably the most acceptable, effective, and safe synthetic compounds that had been developed for this use. The mode of action of the thiazides clearly tied back into the basis on which salt restriction had been used previously in the management of hypertension. They could be used with other antihypertensive agents making the net effect greater or permitting the use of smaller amounts of the other agents to gain a desired effect. Their introduction in the middle of the late 1950's coincided with the discovery of other drugs that extended the range of treatment from mild to severe hypertension.

These important achievements have brought about a treatment for hypertension that is accepted by physician and patient. The effectiveness of these new drugs in reducing the major complications of hypertension is attested by long-term studies by the Veterans Administration. The advances in treatment have increased efforts to gain a better understanding of the origin of the disease, its manifestations and consequences. A hypertensive or edematous patient may count himself fortunate because the newer methods of treatment not only extend his life but also relieve him of many of the complications that earlier treatments produced.

Other advances in the study of kidney problems have been important for the betterment of the health and welfare of mankind. Two that stand out are the development of the artificial kidney as a practical means of treatment, and transplantation of the human kidney (discussed above). In the majority of cases, patients can return to useful productive well-being for many years and in some cases indefinitely, patients who without treatment would have died within months.

RENAL DISEASE

In the early 1940's a number of European investigators developed primitive models of artificial kidneys, designed to remove from the blood of patients with kidney failure the uremic poisons that eventually would cause their death. One of these devices was introduced into the United States and somewhat modified. The changes in design made this instrument a practical tool for the treatment of patients with kidney failure. It was first used to treat patients with acute kidney failure to tide them over the episode of uremia (excess urea in the blood) until their own kidneys began to function again. With emphasis on the use of the artificial kidney as a means of treatment, considerable experience was also gained with other methods of treating acute kidney failure. Knowledge gained from battle casualties and civilian casualties vastly improved this treatment.

Because tubes had to be inserted surgically into veins and arteries each time, repeated treatments with the artificial kidney over an extended period were impractical. In 1960, however, a special plastic tube was developed through which blood could be shunted between artery and vein in the forearm; this made possible repeated access to the blood stream and also repeated and chronic treatment with the artificial kidney. It thus became possible to keep patients with total renal failure, indeed, patients without any kidneys at all, alive and well by repeated treatments with the artificial kidney that cleaned their blood of wastes.

In 1964 an artificial kidney was placed in a home and the patient was treated there, first by a nurse and later by the patient's wife, after she had been trained. This step led to the development of "home dialysis," a form of treatment that allows a patient to treat himself at much less cost than in the hospital and with a considerably more flexible schedule. Smaller, more effective, and less expensive artificial kidneys have been developed. Within a few years, a patient with two medium size suitcases may take along on trips all of the components of an artificial kidney needed to treat himself.

The development of a simple technique for the temporary, and occasionally permanent, treatment of uremia has been made possible by developments in the use of peritoneal dialysis. This technique, first used in the early 1950's, consists of putting fluid into the peritoneal or abdominal cavity, through a small puncture wound in the abdominal wall. The fluid is allowed to remain there for a short period of time and then removed. The toxins of uremia pass from the blood into this fluid across the peritoneal membrane, which lines the abdominal cavity, and so are removed when the fluid is drained off. Improvements in the technique and the use of plastic materials make this an extremely simple, practical, and effective method for the short-term treatment of kidney failure in even smaller community hospitals.

The ability to keep patients with terminal renal failure alive with the artificial kidney prompted the first attempts at transplantation of the human kidney. At first, kidneys, taken from persons who had just died in the hospital, were transplanted into patients with terminal uremia who were being maintained on the artificial kidney. Seven such cases were reported in 1953. In three of them, survival was considerably longer than predicted from animal experiments. In 1954 the first successful kidney transplant between identical twins was accomplished. A patient terminally ill with kidney disease and with severe high blood pressure became completely normal by the transplantation of a healthy kidney from the identical twin and the removal of the patient's diseased kidneys.

The major problem in kidney transplantation is the immunological difference between tissues of two individuals, a difference which is less in the case of identical twins. The problem of modifying the immune reaction against the implanted kidney without totally destroying the ability of the patient to respond to other kinds of bacterial and fungal invasion is still a major part of research endeavor in kidney transplantation. Nevertheless, in 1959 successful kidney transplantation was accomplished between individuals who were not identical twins. In this case, radiation was used to suppress the antibody response. Subsequent work demonstrated that an immunosuppressive drug (azothioprine) could also modify the immune response in the human and this plus cortisone have been the mainstays of renal transplantations since that time.

It is now possible to match tissues between related individuals and predict the outcome of renal transplantation. If the tissue match is excellent between a brother and sister pair, one can predict that the 2-year survi-

val, and probably the 10-year survival, will be well
over 90%. In the case of a kidney transplanted from a
just deceased person, survival is not quite as good, only
60–70% for a 2-year period. Nevertheless, a considerable number of individuals now are alive and well,
without signs of renal disease, after more than a 5-year
period following the transplant.

The necessity of maintaining chronically ill patients
with an artificial kidney prior to transplantation has
required team work between the artificial kidney groups
and surgical transplant groups. Central typing agencies
have become available into which information about
both prospective recipients and prospective donors is
fed. Prospective donors are "matched" against the pool
of recipients and kidneys are donated to the best
matched recipient. Improvements in preserving kidneys now permit organs to be sent by air across country
to the ideal recipient.

In addition to the therapeutic applications of kidney
transplantation, studies of the patients who have had
transplants have increased the knowledge of cancer, of
high blood pressure, and of glomerulonephritis (a form
of kidney disease). In the past 20 years, the metabolic
function of the kidney has become better understood
making possible more effective treatment.

The role of the kidney in the retention of salt and
water and the production of edema (dropsy) in heart
failure and liver failure has been studied in some detail; the role of the kidney in the genesis of high blood
pressure has already been discussed.

Studies of the excretion of bacteria by apparently
normal people have shown a relation between the
number of such bacteria and the later development of
kidney disease and even of high blood pressure. Significant developments in the surgical field have contributed to the prevention of subsequent kidney failure
due to infection or obstruction.

An important development has provided surgical
techniques to prevent the reflux of urine from bladder
back to kidneys through the ureters, the tubes through
which urine normally flows from the kidneys into the

bladder. In 1937, an improvement in the surgical technique for correcting obstruction of the kidney at the junction with the ureter was a landmark in treating obstructive kidney disease.

Animal experiments have clearly delineated two kinds of kidney disease. In one kind an antigen–antibody "complex" is deposited in the kidney and this, in many respects, appears to be a prototype of human glomerulonephritis, an inflammation of the kidney tubules. In the second kind, animals may actually form antibody to their own kidneys, and this too has been demonstrated in at least one case of human kidney disease. The interrelationship between the production of renin by the kidney and the production of aldosterone (a steroid) by the adrenal has been a significant advance, leading to a valuable diagnostic test for tumors of the adrenal gland.

Information has accumulated also regarding toxins that may cause kidney disease, such as phenacetin, common in headache remedies; methysergide (Sansert), a substance previously used for the treatment of migraine headaches; mercury; outdated tetracycline; and a number of potent antibiotics. This knowledge has been critical to prevention and even treatment of kidney disease in the human.

PULMONARY DISEASE

Advances in the diagnosis and treatment of disease processes usually are based on new knowledge in the basic disciplines. This is certainly true of pulmonary disorders. Knowledge of the structure of the lung has been enhanced by improvements of the light microscope and particularly by the development of the electron microscope, which reveals details that cannot possibly be seen through the light microscope. An epithelial (cellular) lining has been found in the air sacs (alveoli) of the lungs and a surface active substance (surfactant) on the lining has been identified. Studies of the sizes and the movements of the different cellular parts of the lungs have already contributed to the knowledge of the normal lung and will lead to a better understanding of pulmonary disorders.

Advances in the physiology of respiration have revolutionized the understanding of the basic mechanisms of many lung disorders and have led to new concepts of lung disease. Accomplishments made in the past 25 years include improvements in methods of studying the mechanics of respiration and alveolar ventilation, in techniques for accurately and rapidly determining blood gases, in the assessment of oxygen and carbon dioxide diffusion across the lung membranes, and in the determination of irregular perfusion of the air sacs.

The development of artificial atmospheres for use in space cabins has provided information on the toxic effects of oxygen relative to pulmonary lesions, as in the lungs of infants kept too long in incubators, and of cardiac patients with pulmonary lesions kept too long in oxygen tents. Research on the performance of humans in sealed environments, relative to ocean floor and outer space, led to ideas and techniques applicable to problems of pulmonary disease, including improved treatment of disease states that impair normal lung function. Such advances have led to new concepts requiring new terminology. Pulmonary disorders are now described in terms such as hypoventilation (low) or hyperventilation (high), hypoperfusion, hypercapnia (excess carbon dioxide) and hypocapnia (abnormally low carbon dioxide), hypoxia (too little oxygen), and respiratory acidosis or alkalosis.

A physician treating pulmonary disease must be familiar with pulmonary anatomy, physiology, biochemistry, pharmacology, and microbiology if he is to understand the mechanisms involved in dyspnea (difficult breathing), cyanosis (blood discoloration due to too little oxygen), and respiratory failure. He must realize what is going on in the lungs and circulation when he sees patients with such disorders as bronchial obstruction, collapse of the air sacs, pulmonary consolidation, and pulmonary infarction (tissue death due to blockage of blood circulation).

Today the pulmonary physician has access to such knowledge and is aided by many new diagnostic procedures. Refined physiological techniques enable the physician to distinguish, for example, between obstructive and restrictive pulmonary disorders. He can obtain pulmonary secretion for microbiological or cytological examination. He can now obtain lung secretions by special methods of inducing sputum production, by insertion of a tracheal catheter or tube, by bronchial brushing, or by needle biopsy through the thorax. Visualization of the interior of the tracheal and bronchial tubes has been improved by new and better telescopes and more recently by the development of the

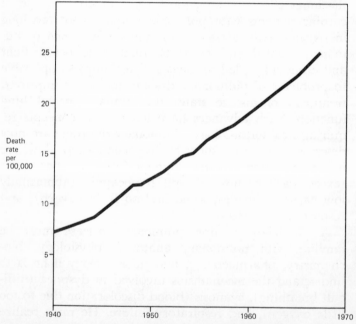

FIG. 7. Age adjusted death rate per 100,000 population from malignant neoplasms of the respiratory system since 1940. *Source:* Office of Health Statistics Analysis, National Center for Health Statistics, Health Services and Mental Health Administration, Public Health Service, Department of Health, Education, and Welfare (1971).

fiberbronchoscope (a flexible glass-fiber viewing instrument). Photographs of these interiors can now be made. Transbronchial biopsy can now be done instead of the more drastic open-lung biopsy in certain diseases. Techniques of lung scanning, pulmonary angiography (X-ray pictures of pulmonary blood vessels), and cardiac catheterization can now be used in disorders of the pulmonary circulation. Sophisticated radiologic techniques developed in recent years can be used as diagnostic tools by the pulmonary physician. Because of an increasing incidence of lung cancer (Fig. 7), associated with cigarette smoking for example, methods of obtaining and examining pulmonary secretions have been an important development for increased knowledge of the pathology of lung disorders.

The treatment of obstructive airway disease and bronchopulmonary pus formation has improved greatly. The chest physician has learned to apply the fundamentals of pulmonary physiology to the treatment of patients. By measuring the volume of lung air before and after administering aerosol to dilate the bronchial tubes he can estimate the reversibility of conditions in bronchitis and emphysema. Improved instrumentation has provided him with better devices, such as special vaporizers with which to administer aerosol bronchodilators or high concentrations of heated mist to liquify semisolid secretions in the air passages, positive pressure devices to assist lung ventilation or introduce medications, and mechanical ventilators that will breathe for the patient temporarily when necessary. Respiratory intensive care areas in hospitals have been developed, with trained personnel, for the handling of particularly difficult cases.

During World War II and thereafter, thoracic surgery and the postoperative management of patients developed tremendously, facilitated by greatly improved anesthesia and antibiotics.

Cigarette smoking has been recognized as a factor in lung cancer and, along with air pollutants, in pulmonary emphysema and chronic bronchitis. Early detection of lung disorders is increasing through tuberculin testing in schools, routine chest X-rays in medical examinations, X-ray surveys in industries and schools, and emphysema screening tests. Tuberculosis remains the major pulmonary problem in the less developed regions of the world.

HEMATOLOGY

Drugs are transported within the body by that great transport system, the blood, the liquid tissue on which the hematologists focus their attention. Whole blood, as it circulates within the vessels, consists of a very complex fluid, called plasma, which carries suspended red blood cells, white blood cells, and blood platelets. Red cells or erythrocytes contain the hemoglobin molecules that carry oxygen to the tissue cells and remove the waste product, carbon dioxide. The white cells (leukocytes), like the other cells of the body, consist of a nucleus embedded in cytoplasm; but neither the red blood cells nor the blood platelets of man contain nuclei. The platelets are tiny bits of cells (cytoplasm) that, under certain conditions such as prevail at a point of physical injury, rapidly change their shape and their biophysical characteristics and help to initiate blood coagulation. This complex stuff, this blood, performs several highly important functions in the body, interference with which causes disorder and disease.

During the past 30 years the discipline of hematology has experienced great activity and has broadened to such an extent that subdisciplines have grown up around the major components of blood, particularly around the red cells, white cells, platelets, and the plasma components involved in blood clotting and immunity.

The red cells must function properly and be present in sufficient numbers; otherwise, one of the blood diseases (called anemias) will occur, diseases in which too little oxygen is carried by the blood (hemoglobin) from lungs to tissues. The red cells are formed in the bone marrow. There is now convincing evidence that the rate of red cell formation is controlled by a special hormone.

Red cell production is stimulated by lack of oxygen
and in anemia plasma proteins, called mucoproteins,
derived mainly from the kidneys, carry to the bone
marrow a signal for an increase in the rate of red cell
formation. Indications have been obtained that certain
refractory anemias, previously difficult to control, may
be benefited now and in the future by treatment with
derivatives of male sex hormones. These advances in
knowledge have led also to a new understanding of the
mechanism, of polycythemias, disorders in which there
is an abnormally high number of red blood cells associated, at least in one variety of this disease, with too
little oxygen in the arterial blood. In some other kinds
of polycythemia, kidney or liver tumors independently
secrete hormones that induce red blood cell production.

Molecular biology has contributed importantly to
hematology and other areas in distinguishing between
normal and abnormal hemoglobin molecules, and the
blood diseases known as hemoglobinopathies are associated with genetically determined alterations in
amino acid synthesis. The information heralds future
advances in the management of sickle cell anemia and
Mediterranean anemia, caused by hereditary abnormalities in the formation of the hemoglobin molecule. In
Mediterranean anemia there is a limited production of
one of the components of the normal hemoglobin
molecule, i.e., of one or the other of its polypeptide
chains.

Advances in understanding nutritional anemia with
the discovery of the biochemical roles of folic acid and
vitamin B_{12} have led to a new appreciation of the roles
of these vitamins and have led to improvements in the
management of pernicious anemia. Improved methods
for measuring the amount of folic acid and vitamin
B_{12} in the blood serum have made it easier to recognize
anemias caused by deficiencies of these vitamins.

The nature and origin of anemias caused by iron
deficiency have become much better understood by
the discovery of factors controlling dietary iron assimilation and of transferrin, a plasma protein that
transfers iron molecules from the blood plasma to the

developing red cells in the bone marrow. Knowledge of the iron deficiency anemias has advanced with the use of radioactive iron.

During the past 3 decades, progress has been made also in understanding the mechanisms of the hemolytic anemias, those anemias that result from a destruction of too many red blood cells in the body. The use of radioactive chromium to determine the survival time of red cells in health and disease has permitted the location of sites, either in spleen or liver, of red cell destruction. These anemias are associated with defects, usually genetic, in the red cells themselves or with an injurious plasma environment that results from certain drugs or from the destruction of red cells by certain antibodies.

There are several types of white cells (leukocytes) in the blood. Studies of the kinetic patterns of these have provided basic information on their normal life-span and sites of origin and destruction in the tissues. Their typical kinetic patterns are now known for the various leukemias, diseases in which an abnormal type of white cell overpopulates the bone marrow and often overloads the blood. It is now known that leukemia may be produced and transmitted in laboratory animals by viruses, a discovery that is of great potential importance to the understanding and possible treatment of leukemia in the human being. A viral cause of human leukemia, however, has not yet been established.

Another important discovery has been that of chronic myelogenous leukemia, which is associated with an abnormally small size of a particular single chromosome and usually with the appearance of an abnormally small amount of a particular chemical compound found in normal white cells. This chromosomal aberration, which is not necessarily genetic, but which is involved in the development of leukemia, also is not necessarily incompatible with the possibility that leukemia may be caused by a virus. The long-term observation of the Hiroshima population and studies of patients who have received irradiation in the treatment of spinal disease, have shown that irradiation can cause

leukemia in man. It is possible that a number of stimuli
may evoke the body response seen as leukemia.

The emergence of electrophoretic methods, methods
by which molecules or particles that carry electrical
charges can be separated in an electric field, have led
to a new understanding of the nature and origins of
normal and abnormal proteins (globulins) in the blood
plasma. The production of certain abnormal proteins
by blood tumor cells (plasma cells) and the presence of
certain unusually large and viscous proteins in the blood
plasma are being studied intensively. Proliferation of
such tumor cells causes abnormality in the production
of gamma-globulin, a plasma protein normally associated with functional antibodies, and has been recognized as the basis of the frequent development of
clinical infections. The tumorous blood cell growths, in
addition to their local effects on bones, decrease the
amount of functional antibodies.

During the past 3 decades there has been a considerable advance in the treatment of leukemia with
drugs. Treatment of childhood leukemia with combinations of drugs can increase the frequency and
duration of remission, and has extended the life of
many afflicted children. The average survival has increased from 2 months to 2 years and a few patients
appear to be "cured." An even more spectacular advance in treatment has been accomplished with "extended field" irradiation of patients with early stages of
Hodgkin's disease, a disease in which there occurs
initially enlarged lymph glands and in which later
abnormal white cells are found in other organs.

Almost all the cells of the body have two main parts,
a cytoplasm surrounding a smaller body, the nucleus,
which carries the genetic elements. But, as previously
mentioned, although the red blood cells of man and
the human blood platelets originate from cells having
nuclei, they do not contain nuclei. The platelets play a
primary role in the process of blood coagulation, which
is abnormal if the platelets are defective or too few in
number. Among the important advances in the past
30 years is one relative to the disease called by the

mouth-filling name of idiopathic thrombocytopenic purpura. This disease of unknown origin involves an abnormally low number of platelets in the blood, causing bleeding within the skin and other body tissues. The disease is associated with and very likely the result of a blood plasma factor (perhaps an antibody) causing platelets to accumulate in the spleen or liver where they are destroyed.

Important for understanding various hemorrhagic processes is information about the role of platelets in blood coagulation. An important sequence of biochemical events begins with an aggregation of platelets at some spot on the inside of a blood vessel wall or at the site of injury to a vessel. The discovery that a number of drugs prevent this platelet aggregation holds promise that drugs may be used to prevent or even to interrupt an undesirable blood clotting process that otherwise may be detrimental or even fatal to the person in which it occurs.

It has been found that at least 13 components of the blood plasma are involved in a virtual "cascade" of enzymatically controlled chemical reactions that result in the formation of a blood clot. Certain hereditary and acquired diseases have been recognized that are manifestations of the absence of one or more of these blood components.

Similarly the details of many of the conditions and substances that oppose blood coagulation have become recognized and described. Heparin, a drug that opposes blood clotting, was discovered nearly a half century ago, but the anticoagulant Coumadin drugs were discovered and have been used for treatment of blood diseases only within the last 3 decades. Although widely used to prevent undesirable blood clotting, their real value in this connection is still not fully determined. More recent times have witnessed the development of thrombolytic substances that break up or dissolve a blood clot. These substances hold promise of improved treatment of thromboembolic disorders, conditions which may be a fatal sequel to surgery or a heart attack. Such a condition results from a blood clot that occurs locally somewhere in the body and is carried by the blood to another site, e.g., the lungs or brain, where it clogs up a blood vessel with serious consequences. Examples of substances that dissolve blood clots are the enzymes streptokinase and urokinase. Although their clinical usefulness has yet to be conclusively demonstrated, these and similar drugs hold promise for the future.

The use of human blood, as whole blood or its main component parts, in the management of human diseases has rapidly increased in the last 3 decades. No substitute has yet been found for human red cell replacement in the patient with deficient oxygen transport from lungs to tissues, nor have practical substitutes been found for other blood cell components. The need for replacement of red blood cells has increased as more radical surgery has been undertaken. Blood plasma and its natural components have not yet been displaced by synthetic substitutes, the so-called volume expanders such as dextran, hydroxyethyl starch, and electrolytic (salt) solutions. Platelet transfusions and the use of gammaglobulin and blood coagulant factors have proved to be useful in certain blood disorders. With an increasing demand for specific blood components, it seems clear that in the future whole blood transfusion will be replaced by transfusion with blood fractions and components. Techniques have been developed and improved during the past 30 years for separation of whole blood into its various cellular and plasma components, for use in transfusions. Clinical practice has benefited greatly from improved methods of blood banking and preservation. Today red blood cells can be frozen and preserved indefinitely.

The circulating blood not only has the emergency function of blood coagulation just discussed but, as mentioned, is also the great river of transport within the body. Among the things it transports, in addition to oxygen, carbon dioxide, foodstuffs, and waste products of the tissues, are the hormones, so important to the regulation of the body processes.

CHAPTER III

Emanating from endocrinology over the last 30 years has been the discovery, purification, structural elucidation, and synthesis of hormones of the endocrine glands of the body, glands that pass their secretions directly into the circulating blood. All aspects of empiricism, unrelated research, and unanticipated discovery have been involved in these achievements. Not infrequently the key unlocking the door to discovery has been observations on individual patients. The availability of new hormones starts an immense number of important developments and applications. Physiological, biochemical, pharmacological, and toxicological studies are performed first in animals and then in a selected number of people. After much effort, and at the end of a developmental sequence, the hormones become the practicing physician's tool.

Hormones have been utilized in medical practice in many important ways other than for the correction of a hormonal deficiency; for example, in the regulation or inhibition of a normal physiological process such as ovulation. Stimulation or suppression of an endocrine organ or organ system is now performed routinely with hormones to test for normal or abnormal function, an essential diagnostic procedure. In a pharmacological sense, hormones administered in therapeutic doses become drugs. An exaggeration of the biological action of hormones or an entirely new biological effect, observed when used in large doses, are reasons why physicians administer hormones. Vigorous and successful attempts have been made to design and synthesize hormone analogues, chemically similar but not identical to the natural hormone, that produce more desirable effects than the natural hormones from which they are derived. In 1953 and 1955 two analogues of the natural hormone, hydrocortisone, were developed, one decreasing and the other increasing the degree of salt excretion by the kidney, as compared with the natural hormone.

The availability to the physician of cortisone and hydrocortisone has improved the health of patients regardless of sex or age. These natural hormones akin to human endocrinology have pervaded every aspect and specialty of medicine as pharmacological agents or drugs for use in nonendocrine disorders. This important development began in 1936 with the isolation of a few milligrams of cortisone and hydrocortisone from adrenal glands, followed by the elucidation of the chemical structure. With only tiny amounts of the hormones available, various physiological activities were demonstrated although the therapeutic importance of these hormones was not realized at the time. The objective in 1942 was to obtain sufficient supplies of cortisone to examine the possible application to medicine. The attainment of this goal was an outstanding success brought about by the imagination, talent, and cooperation of chemists who, in groups and as individuals, worked together to accomplish first the partial and then the total synthesis of cortisone.

Clinical studies were begun in 1949. Because of the many physiological effects of cortisone, even within the area of endocrinology, it is difficult to generalize the role this hormone has played in the treatment of patients. One unique and specific use is to maintain health and longevity of patients who have no adrenal or pituitary glands, a goal not previously possible. With the availability of cortisone, removal of the pituitary and adrenal glands for a medical reason is performed frequently.

Equally dramatic is the administration of cortisone or hydrocortisone to inhibit the activity of the pituitary–adrenal glands of children or infants and diminish excessive secretion of other adrenal hormones. In these patients, the inability to synthesize and secrete normal quantities of hydrocortisone results in an excess secretion of other adrenal hormones that cause adverse effects. The adverse effects of adrenal abnormalities are dramatic. In some young patients so little hydrocortisone is produced by the adrenal glands that the child will die without the administration of cortisone or hydrocortisone. Girls may become masculinized and boys precociously developed sexually. Some children experience hypertension. Early diagnosis and daily treatment with hydrocortisone can prevent these changes and save lives.

Identification in 1953 of the adrenal steroid, aldosterone, that controls sodium and potassium metabolism introduced another series of events. Methods for measuring the concentration of this hormone were developed. The unique and complex system involving renin and angiotensin that regulates release of the hormone was described and the components of the regulatory system became measurable. The metabolic effects of aldosterone and its levels in normal and sick people were established. In 1955 another cause of hypertension had been detected. A patient with a tumor of the adrenal gland was secreting too much aldosterone. Removal of the tumor reduced the secretion and cured the hypertension. Today this diagnosis and treatment are not unusual. The clinical and laboratory criteria for diagnosis of the disorder are found in textbooks of clinical medicine. With the discovery of a hormone an antagonist, an agent which blocks the effect, can be developed. An antagonist to aldosterone was developed several years after the discovery of that hormone and is now available to physicians. The antagonist is administered to counteract electrolyte or salt effects of aldosterone on the kidneys of patients with various types of edematous disorders.

At least 40 other steroids similar to cortisone, hydrocortisone, or aldosterone have now been isolated from the adrenal gland, but none have been as important as cortisone and hydrocortisone. Because of the multiple actions of cortisone and hydrocortisone, some good and some bad, the clinician has had to learn to use them wisely.

Other accomplishments associated with the cortisone epic are the synthesis of steroid hormones, now utilized to inhibit fertility, and the anabolic steroids, which facilitate the assimilation of nutritive substances and their conversion into living tissue. The anabolic hormones are utilized to improve the health of debilitated patients. The developments are important in controlling population and in the aging problems of older people.

The success of the antifertility hormones, known best

as "the pill," is well established. Few people are aware,
however, of the long history of basic research and industrial investigation that led to the development of the
birth control pill (Fig. 8).

Because of multiple biological effects of steroids other
than the inhibition of fertility, the control of fertility is a
complicated problem. Utilization of the antifertility pill
is an example of a physician administering medication
for a prolonged period to a healthy person to regulate a
normal physiological process. This situation under-
scores the necessity of understanding more about the
health and physiology of normal people.

Other achievements have been made in the diagnosis
and treatment of patients with a tumor of the adrenal
glands that causes these glands to secrete abnormal
amounts of hormone, as in the case of hypertension
mentioned above. Newly developed blocking agents
have been used beneficially both preoperatively and
during operations on patients with this tumor. Similarly,
advances have been made in the diagnosis and treatment
of other related tumors.

At the base of the brain is an endocrine gland, the
pituitary, consisting of an anterior lobe and a posterior
lobe that secrete a number of hormones. Physiological
activities of seven anterior pituitary gland hormones
were discovered more than 30 years ago and since then
each has been isolated in highly purified form. The
chemical structures of some pituitary hormones have
been elucidated and a few have been synthesized. In
the anterior lobe of the human pituitary gland there is a
relatively large amount of a particular hormone called
growth hormone since it stimulates body growth. Fortunately this hormone resists immediate destruction after
death, for at present the body of a deceased person is
the only source of this hormone for use with human patients. The growth hormone is species specific; that is,
growth hormone from a nonprimate source, such as the
ox or pig, is not active when administered to man. Administration of purified growth hormone to children
deficient in this hormone dramatically stimulates body
growth. The chemical structure of human growth hor-

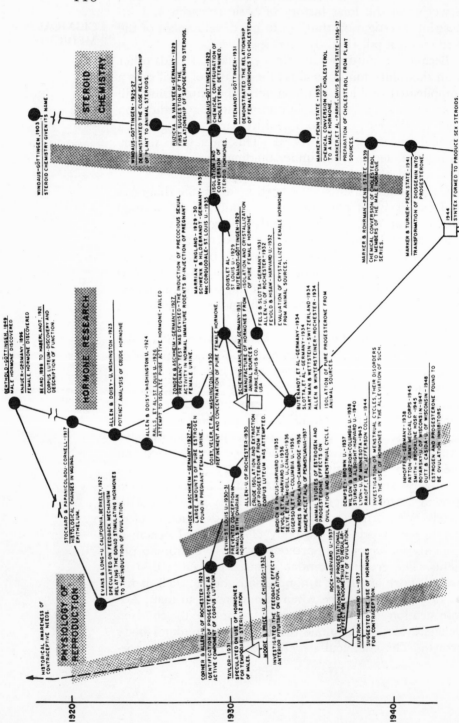

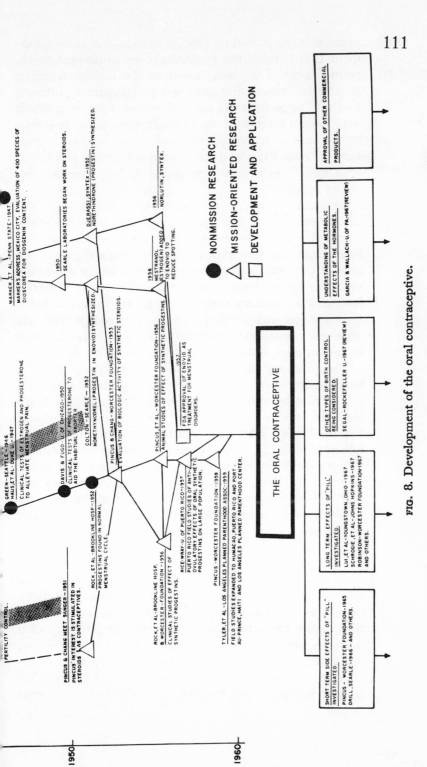

FIG. 8. Development of the oral contraceptive.

Adapted from Volume 1, "Technology in Retrospect And Critical Events in Science," prepared for the
National Science Foundation by the Illinois Institute of Technology Research Institute under Contract NSF-C535

mone was elucidated in 1966, a tremendous accomplishment that heralds the synthesis of an important therapeutic agent. The great variety of the biological effects of human growth hormone on carbohydrate, protein, and fat, as well as on mineral metabolism, suggests it may have a large number of therapeutic uses, as is true in the case of hydrocortisone or cortisone. The importance of knowledge and specialized training of physicians who utilize these complex agents to help the sick becomes apparent.

Elucidation of one of the closer links between the brain and endocrine system has been methodically developing over the last 40 years. The smaller posterior lobe of the pituitary gland develops from a region in the base of the brain called the hypothalamus, to which the posterior lobe remains attached. The larger anterior lobe of the pituitary does not originate from the brain but connects with the hypothalamus by special vascular channels. Hormones secreted from the hypothalamus to the anterior pituitary gland (lobe) control the release of the pituitary hormones. Evidence indicates that for each hormone of the anterior pituitary gland there is a companion hypothalamic hormone that regulates its release. In 1966 a highly purified preparation of a natural hypothalamic hormone called thyrotropin-releasing hormone (TRH), which causes release of the hormone thyrotropin from the anterior pituitary, was found active in man. By 1969, the chemical structure of TRH was determined, the hormone was synthesized and the synthetic hormone was found to be active. This is the first chemical structure of a hypothalamic-releasing hormone to be elucidated. The type and pattern of response of the body to natural and to synthetic TRH are the same.

Hypothalamic hormones that control release of other hormones of the anterior lobe of the pituitary also have been purified. The LH-RH is a hypothalamic hormone that causes the release from the anterior pituitary of a hormone that acts on the gonads. When this hormone was given to adult women it caused a secretion of increased amounts of luteinizing hormone, a hormone important in the release of an egg from the ovary. Current research suggests that egg production in the human female and fertility in the male could be more precisely controlled with LH-RH. A possible future development is synthesis of hypothalamic-releasing hormone analogues that will be prepared and used as inhibitors of the natural hormones. An analogue of the hypothalamic hormone, LH-RH, because of its specificity of action, could become the antifertility agent of choice. With availability of the hypothalamic hormone that controls the release of growth hormone from the pituitary gland, the rate of body growth could be stimulated in selected patients.

In 1928 evidence indicated that two hormones must be present in the posterior lobe of the pituitary gland and one, oxytocin, was isolated in 1949. By 1953 oxytocin and the other hormone, vasopressin, were synthesized. These were the first biologically active peptide (protein derivative) hormones to be synthesized. After removal or destruction of the smaller posterior lobe of the pituitary gland, patients pass 7–10 quarts of urine/day due to the inability of the kidney to concentrate urine, a disorder called diabetes insipidus. Administration of vasopressin corrects urine flow in patients deficient in this hormone and oxytocin is used for induction of labor in pregnancy, control of hemorrhage at childbirth, correction of flaccidity of the uterus after childbirth, and improvement of milk flow during lactation.

Forty years ago, surgical removal of the thyroid glands was the only possible treatment for thyroid disease, one of the most common endocrine disorders. Since that time, advances in the study of thyroid disease have included measurement of serum levels of thyroid hormone, thyroid antibodies, and thyrotropin, and newly devised thyroid function tests using radioactive iodine. Development and application of radioactive iodine for medical use, initiated in 1941, has been an unequivocal success. With trace amounts of this radioactive isotope and with properly trained people and appropriate instrumentation to detect the radioactivity, malfunction of the thyroid can be readily diagnosed. The treatments of patients with radioactive iodine to decrease the activity of a toxic goiter or with thyroid hormone to decrease the size of a nontoxic goiter are very valuable types of thyroid treatment that have been developed. When the thyroid gland is underactive, the synthetic thyroid hormones, thyroxine or triiodothyronine, can be administered singly or in combination. Thyroxine was discovered in 1915 and triiodothyronine in 1950. Triiodothyronine has been shown to possess a biological potency greater than that of thyroxine by most methods of assay; however, it is normally present in the serum and thyroid glands of humans in considerably lower amounts than is thyroxine.

Antithyroid compounds, first developed in 1940 before radioiodine was available, are still the most commonly used agents for treating overactivity of the thyroid. Without them a great deal of morbidity as well as mortality would result from this disorder, as can be discerned from earlier medical records.

Thyroid cancer does occur, but its incidence is much less than previously thought. Physicians now have many tests to detect its presence. When thyroid cancer has spread in the body, the administration of radioiodine has been lifesaving to selected patients.

A new hormone which has been wending its way into the physician's armamentarium since 1961 is thyrocalcitonin, which is secreted from the thyroid and lowers serum calcium levels. The chemical structure of human thyrocalcitonin is known; it has been synthesized and the synthetic compound is active in man. Patients with hypersecretion of this hormone have been found and undoubtedly others with hyposecretion soon will be discovered. The therapeutic value of thyrocalcitonin in a disease causing deformation of the bone has been demonstrated. Among the potential uses of thyrocalcitonin currently under intensive study are the decrease of elevated calcium levels, and the prevention or repair of damage from breakdown of the bones of the skeleton that occurs in the elderly and in women after the menopause.

A dream of patients with diabetes mellitus has been to take diabetic tablets (oral hypoglycemics) rather than injections of insulin to control their blood sugar levels. In 1942 some physicians treating typhoid fever patients with an experimental sulfonamide observed a lowering of blood sugar levels. By 1946 drugs with chemical structures of sulfonamides or containing the sulfonylurea radical that had been found to lower blood sugar levels in animals were administered to patients, and were found to have the same activity. Large numbers of patients with diabetes mellitus the world over are being treated with one of the several sulfonylurea agents, "diabetic tablets," now available. The results of some recent studies, however, have raised a question as to the actual effectiveness and value of the oral hypoglycemics.

The elucidation of chemical structure of insulin, a hormone secreted by the pancreas, was announced in 1955 and insulin was synthesized in 1964, a phenomenal achievement. In 1969, circulating insulin was found to exist in two distinct forms, "big" and "little" insulin. Another hormone of the pancreas, glucagon, which raises rather than lowers blood sugar levels, was purified in 1955 and its chemical structure known by 1957. Its role in health and disease of man is under intensive study. Physicians have already been administering glucagon to raise the blood sugar levels of patients who have taken an overdose of insulin.

A new family of hormones with at least 16 naturally occurring forms, the prostaglandins, is another possible new therapeutic tool. Although the existence of this group of hormones has been known since 1930 and although it was observed in the 1930's that these hormones would lower blood pressure, research was limited until after World War II, perhaps because they were available only in very small amounts. By 1964 a major source for these hormones was found to be the seminal glands of sheep. They are present in high concentration in male semen and have been found in the blood of pregnant women during labor. Prostaglandins have been found not only to lower blood pressure but also to induce labor when injected in women between the 34th and 44th week of pregnancy. This family of hormones is now being studied as a possible antifertility agent, as an agent to induce labor at term, as an agent to decrease gastric secretions, and as a vasoconstrictor in nasal tissue. If and when prostaglandins become available for use in clinical practice, surgical abortion may become a thing of the past. Such seemingly unrelated actions of hormones are discovered frequently and often exciting and unexpected multiple therapeutic uses are found. This seems to be especially true of the prostaglandins that are now being widely studied and that show promise of becoming important in hormonal therapy in the next decade.

Endocrine therapy of cancer of the breast and prostate is well established. Removal of the pituitary, adrenals, and gonads is performed to help patients with metastatic cancer of the breast or prostate. Pharmacological doses of female sex hormones (estrogens), male sex hormones (androgens) or analogues of estrogens and androgens are important agents to be used at this stage of these diseases. Although the administration of hormones or the removal of endocrine organs fails to cure cancer, temporary dramatic disappearance or decrease in the size of the cancer occurs in some patients.

The availability of hormones has helped physicians in many ways. By knowing the chemical structures or the specific chemical properties of hormones as well as having hormones available as reference standards, specific and relatively rapid and inexpensive methods have been developed for the accurate measurement of these substances. Measurement can now be made of unimaginably minute amounts of hormones present in small samples of human serum or urine, amounts even as minuscule as one trillionth of a gram.

With these highly sophisticated micromethods, physicians of every specialty have learned about the normal and the abnormal secretion of hormones. For instance, not infrequently benign or malignant tumors of nonendocrine organs of the body, such as the lungs, stomach, and liver, have been found to secrete small amounts of one or two of the natural hormones. The hormones involved in the normal menstrual cycle have been measured in great detail. Present evidence indi-

cates that the many hormonal changes measurable in
obese patients are more likely to be the result rather than
the cause of the obesity.

The many endocrinological discoveries and applications during this 30-year period have been confirmed, reconfirmed, and extended, and have set off chain reactions of advance in both clinical and basic science. Each step can be clearly delineated. The developments in the past 30 years demonstrate that factors and forces ordering these steps often seem of a heterogeneous, empirical, and unplanned nature. Often they have been more akin to the esthetics, emotions, and philosophy of man than to arithmetic and orderly logic. Man's drive to learn concisely and specifically may be the most important ingredient of success. As long as man exists, the problems of the sick people will exist, but with new knowledge physicians will be able to improve human health.

The endocrinological system and its multiplicity of interrelated hormones has the greatest significance for a well-regulated functioning of the healthy body. But another system of control is of the utmost importance to the precisely timed unfolding of the various parts of the developing organism—the genetic basis of life itself—the genetic code within the cells of the body. When something goes wrong with the precise genetic control of this complex development, inherited disease usually appears.

MEDICAL GENETICS

As a health discipline, medical genetics has grown to its present youthful stage since 1940. It is that branch of medicine that relates genetics to health and disease and is concerned with hereditary disorders. As in other branches of medicine, this involves diagnosis, prognosis, treatment, and prevention. Progress in better understanding of the scientific basis of medical genetics has permitted its development as a legitimate medical discipline.

In 1940 human genetics, including medical genetics, was largely an area of pedigree studies involving statisti-

118

CHAPTER III

cal techniques that tried to compensate for small family size, inability to make experimental human matings, and other research limitations of man considered as a subject for genetic study. Scarcely more than two traits, color blindness and the ABO blood groups, groups identified by clumping reactions of red blood cells when brought into contact with blood sera containing various antibodies, had been included in human population studies. More than 30 years earlier the concept of inborn errors of metabolism had been formulated but the significance of this either to the practice of medicine or to genetic theory had not been recognized. The precise enzyme defect in any hereditary disorder of man remained a mystery. Even the correct number of chromosomes in normal man was not known.

Progress in genetics has been mostly in the areas of biochemical genetics and genetics of the cell, but advances have occurred in the inheritance of different immune responses and in understanding the classification and natural history of genetic diseases. Pharmacogenetics, the study of inherited variations in response to drugs, is a development of the last 2 decades.

About 3 decades ago, the concept of "one gene–one enzyme" was developed from studies with a fungus called *Neurospora*. This concept also was implicit in an earlier study of alkaptonuria in man, a metabolic defect resulting in the excretion of certain chemicals (alkapton bodies) that cause the urine to turn dark on standing. A structural abnormality of the hemoglobin molecule was associated in the late 1940's with sickle cell anemia, a serious inherited disease. The understanding of this disease was extended by the discovery that sickle cell hemoglobin and a normal hemoglobin differed from each other only by one amino acid group. The concept of "one gene–one polypeptide chain" (chain of amino acids linked together) evolved a refinement of the "one gene—one enzyme" concept.

Enzyme deficiencies, the basis of inborn errors of metabolism, were identified in the early 1950's. One of these diseases, resulting from inborn errors of carbohydrate metabolism, causes low blood sugar and is

sometimes accompanied by convulsions. The liver becomes engorged with glycogen and fat, resulting in
death in infancy. Another inborn error of metabolism
is a congenital deficiency that causes an amino acid,
phenylalanine, to accumulate in the blood. As a consequence, another acid is excreted in the urine and the
phenylalanine and its metabolites accumulate, resulting
in brain damage and severe mental retardation. The
brain damage can be prevented by keeping the diet low
in phenylalanine, a treatment that is a logical extension
of our understanding of inherited chemical errors of
human metabolism. Specific enzyme changes have now
been identified in over a hundred inborn metabolic
errors of man. Specific dietary treatment, or therapy
utilizing substances necessary for enzyme activity, has
been effective in some diseases, for example, in maple
syrup urine disease, galactosemia, and homocystinuria.
In maple syrup urine disease, urine smells like maple
syrup. It is a fatal disease of the nervous system. Galactosemia results from an error in the metabolism of the
sugar, galactose, a disease that begins in infancy and
causes mental and physical retardation. Homocystinuria
is an inherited error of metabolism in which the amino
acid, homocystine, is excreted in the urine and mental
retardation and other abnormal conditions occur.
Screening methods have been developed which can be
applied to blood or urine samples from large numbers of
individuals, especially newborns, for detection of treat-
able inborn errors. Even prenatal diagnosis by study of
amniotic fluid or cells obtained by aspiration of amniotic
fluid from the mother has been possible in more than 50
diseases.

Another advance in biochemical genetics occurred
with the discovery of hidden variations in both enzymatic and nonenzymatic proteins in man, a condition
referred to as polymorphism. Development of such
methods for studying the physical properties of proteins
as electrophoresis (the separation of certain molecules or
particles in an electric field) led to this advance. In
almost all polymorphisms the factors responsible for the
relatively high frequency of the genes for these protein

variations have yet to be established. A relation to disease susceptibility is suspected.

An aspect of genetics of significance to clinical medicine is pharmacogenetics, the study of variations of responses to drugs caused by inherited differences in metabolism. Pharmacogenetics began with the discovery of susceptibility to hemolytic anemia resulting from the use of primaquine and certain other drugs by persons subsequently found to have a deficiency of a certain enzyme in their red blood cells. It was followed by the observation that some individuals develop prolonged muscle paralysis from succinylcholine, a muscle relaxing drug used as an adjunct in anesthesia. Such persons have a deficiency of an enzyme which normally breaks down this drug. Resistance to the anticoagulant effects of coumarin is another example of a genetic peculiarity made evident in response to drugs.

Study of the chromosomes of man, an aspect of morphology, was greatly facilitated when, in 1956, the correct number of chromosomes in human body cells was shown to be 46. The first recognition that a chromosome abnormality produced mongolism was announced in January 1959. In the past 12 years, chromosome abnormalities, either structural or numerical, have been related to at least six types of clinical disorders: 1) to abnormalities of sex, as in one disease of sexual infantilism and dwarfism, and in another disease in which males are to a variable degree eunuchs; 2) to mental retardation (e.g., mongolism); 3) to complex malformation syndromes, as one which involves various anatomical malformations and mental retardation in children; 4) to certain behavioral abnormalities; 5) to 20–30% of the cases of spontaneous abortion; and 6) to certain tumorous growths as in chronic myeloid leukemia.

Other chromosomal disorders are now recognized, such as 1) genetically different tissues or cells associated together in an organism (mosaicism), 2) abnormalities due to breakage of chromosomes, 3) abnormalities due to translocation of segments of one abnormally broken chromosome to another, and 4) abnormalities due to chromosomal segments that get back into place hind part before. Increased chromosome breakage has been demonstrated in certain inherited inborn errors of metabolism, especially in a particular anemia that may involve dwarfism and mental retardation. The role of parental age in determining the frequency of chromosome aberrations has been established in some instances. Discovery of the basis for familial chromosomal aberrations, recognition of the carrier state (in the form of balanced chromosomal translocation), and prenatal diagnosis by study of the chromosomes of amniotic cells, have important practical applications in a small but significant group of cases.

Developments relative to the genetics of immunological reactions have been along two lines: *1*) the genetic control of the normal immune mechanisms (elucidated mainly through the study of genetic defects in these mechanisms) and *2*) the genetics of tissue compatibility, that is, different tissues sufficiently similar to permit transplantation of organs or tissues from one person to another. The genetics of tissue compatibility has important practical implications not only to organ transplantation but probably also to natural resistance to malignancy.

Advances in the classification of genetic disease, improved delineation of clinical disorders, and better understanding of the natural history of genetic disorders have provided a better basis from which to search for fundamental faults and for effective therapy, genetic counseling, and prevention. The developments in genetics are important to advancements in pediatrics, an area which relates closely to many other medical areas, outstandingly so with respect to inherited diseases and disorders, which often strike in childhood.

PEDIATRICS

The advances in pediatrics have been made by pediatricians, by surgeons working in a pediatric setting, and by microbiologists and pathologists. Advances in the biological sciences, such as chemical agents for use against infectious organisms, have had dramatic impact in pediatrics.

Molecular biology, stimulated by discovery of the basis of sickle cell anemia (a severe hemolytic anemia affecting Negroes in which the red cell hemoglobin molecule differs from normal hemoglobin), demonstrated that the laws of genetics operate through control of the type of protein molecules synthesized by the cells. Since the sickle cell hemoglobin discovery, many abnormal hemoglobins have been detected in various ethnic groups. Diseases involving abnormal hemoglobins are called hemoglobinopathies. The structure of the hemoglobin molecule has been worked out in great detail. A single hemoglobin molecule contains two pairs of polypeptide chains (chains of amino acids). The chains have been separated and put together again in the laboratory. More is known about how molecular structure and physiological properties are related in the hemoglobin molecule than perhaps for any other protein. Abnormalities in the structure of polypeptide chains or an imbalance in their rate of synthesis usually give rise to severe hematological disturbances.

There have been tremendous advances in understanding metabolism and its enzymatic regulation since the concept of inborn errors of metabolism was introduced in 1908. Development of techniques of chromatography, the separation of materials by selective adsorption in columns of a suitable adsorbing substance, have stimulated many studies on inborn errors of metabolism that cause disease. It has been estimated that in the human body there are more than 100,000 different protein molecules, the majority of which are enzymes that regulate metabolic processes, any of which could be defective. Hereditary diseases are being found in increasing number to result from mutations that cause the formation of abnormal proteins (enzymes), the mutations being changes in the character of a gene that are passed on to later generations. Many abnormal conditions can be diagnosed by chromatographic study of the urine or biochemical studies of the blood. A striking example of this is phenylketonuria, the disease that accounts for about 5% of all retardation cases in mental institutions. Routine screening for this defect is now being carried out on most babies born in this country. Immediate placement of the child on a relatively phenylalanine-free diet will result in the physical growth and mental function of that child progressing almost normally.

MEDICINE

Another inborn error of metabolism is congenital adrenal hyperplasia, in which the adrenal gland secretes an excessive amount of an abnormal hormone (cortical hormone) associated with excessive salt excretion and premature virilization. Male infants are born with large genitals and female infants have ambiguous genitalia and are often mistaken for boys. When cortisone became available, treatment with cortisone stopped the abnormal virilization and restored normal salt regulation, enabling the children to grow as normal individuals, even permitting some of these females to have children of their own.

Techniques developed by biologists for the study of chromosomes, now applied to human cells in clinical investigations and diagnostic practice all over the world, have opened up new vistas for application of genetic knowledge to human welfare, a development of great importance in pediatrics. Mongolism, a congenital disease that usually can be recognized in the first week of life, constitutes the largest single group in institutions for the mentally retarded. The normal number of chromosomes in a cell is 23 pairs, or 46. The cells of mongols usually have 47 chromosomes. Another type of mongolism has been found to be associated with an extra amount of chromosomal material and a normal number of chromosomes. Several other chromosomal disarrangements have been found in babies with multiple congenital defects. A major effort is under way to map the human chromosomes in order to locate precisely the genes in each chromosome and correlate them with the observed congenital defects. Eventually important discoveries will come from this long and laborious process.

The applications of biochemistry and cytology are still being developed. One application is the intrauterine diagnosis of genetic diseases through the study of amniotic fluid obtained by aspiration from the mother, and of cells therein derived from the skin of the fetus. The diagnosis of inherited diseases early in pregnancy opens up the possibility of a termination of pregnancy when the fetus is defective. Another application is the screening of children for metabolic errors that are detectable before overt disease symptoms appear, especially children of individuals with inherited disease. As laboratory procedures become increasingly automated, early screening programs become general and preventive genetics becomes a reality. There is a change in attitude by the public from one of passive acceptance to one of aggressive response to inherited diseases; that is, to an attitude of creating the fullest possible life for defectives. The new attitude has had a tremendous effect on the treatment of handicapped and mentally retarded children. The better treatment provides an outstanding example of social benefit arising from basic research in biology.

The discovery that the usually fatal course of acute leukemia in children could be stopped and full, though temporary, clinical remission achieved by the use of chemical antagonists to folic acid ushered in the era of chemotherapy for the control of tumors of many types. Research by teams of scientists and physicians will lead to additional achievements.

No stage in human life has changes of such magnitude as the 9 months spent by the fetus inside the mother's uterus between fertilization and birth. There have been three outstanding achievements in the study of this period; these are with respect to erythroblastosis fetalis, congenital infections, and intrauterine diagnosis.

Erythroblastosis fetalis is a serious hemolytic disease of the newborn with high mortality that has been shown to result from destruction in the fetus of Rh-positive red cells by antibodies from the blood of an Rh-negative mother; these maternal antibodies are formed as a reaction against Rh-positive red cells that entered the mother's circulation, primarily when the placenta is disrupted during separation at the time of birth. A practical program of postnatal diagnosis and treatment for this disease has been developed, including a system of exchange transfusion, a treatment that has saved countless lives and protected many thousands from developing

cerebral palsy and deafness. Severely affected babies
died prior to birth before they could benefit from exchange transfusion. A few years ago, techniques were
developed for prenatal intrauterine diagnosis and transfusion. More recently, it was discovered that the disease
could be prevented by administering small amounts of
concentrated gamma-globulin containing a high level of
appropriate antibody to an Rh-negative mother at the
time of the delivery of her first Rh-positive baby. The
treatment prevents sensitization of the mother by rapidly
removing red blood cells derived from her Rh-positive
baby. In a period of 40 years, a prevalent, serious disease
has been recognized, its causes understood, and effective
techniques for treatment and prevention developed.
This crippling and fatal disease may soon become extinct.

The control of infectious diseases has changed pediatrics almost beyond recognition during the past 30–40
years. Many diseases including malaria, pneumonia,
meningitis, rheumatic fever, tuberculosis, diphtheria,
whooping cough, and other infectious diseases and disorders have been virtually eliminated. New findings of
great potential importance to pediatrics have changed
concepts of the relationship between viruses and human
tissues.

Important advances have occurred in understanding
acquired congenital defects caused by noxious agents
affecting the fetus, particularly during the early months
of pregnancy when the fetal organs and organ systems
are undergoing embryological development. Early
biological experiments in the production of anatomical
deformations (production of monsters) by means of
drugs or diet manipulation during pregnancy, plus the
later human counterpart in the case of German measles,
and the recent tragic epidemic of severe malformations
caused by the drug thalidomide have alerted the pedia-
trician and the public to the unexpected dangers lurking
in pregnancy.

Nutritional diseases of children continue to be serious.
In the United States the commonest nutritional disease
is iron-deficiency anemia. Chronic malnutrition among
Western children, if not the result of socioeconomic

1940

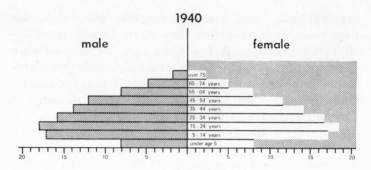

1970

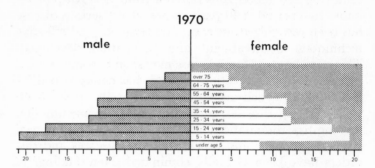

FIG. 9. Percentage distribution of the United States population by age and sex, 1940 and 1970. *Source:* Office of Health Statistics Analysis, National Center for Health Statistics, Health Services and Mental Health Administration, Public Health Service, Department of Health, Education, and Welfare (1971).

situations or lack of education, now most often results from some genetic peculiarity of the child. Pediatric surgeons have developed a technique, the "life line," for handling severe nutritional disorders in young children, a method of long-term intravenous feeding through a plastic catheter passed through a vein into the right side of the heart, enabling the child to grow and gain weight for a given period without taking food by mouth.

All of these accomplishments benefit not only the child but also the young, mature, or aged person he will become. With the reduction of deaths from infectious diseases, medical attention has become focused more and more on old people who now form an increased percentage of the population (Fig. 9). There is also the problem of proper care of the women who give birth to children.

OBSTETRICS AND GYNECOLOGY

In 1940 the maternal mortality rate in the United States was almost 400/100,000 live births (Fig. 10). In 1969, this country achieved an impressive decrease in rate to a low of 27/100,000 live births. Three major advances contributed to this improvement. First was the development of antibiotic therapy. Penicillin, chloromycetin, and streptomycin became major factors in preventing maternal deaths due to childbirth fever caused by infection. The second major advance was the development of blood banks and the refinement of blood

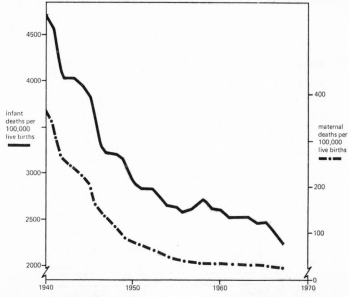

FIG. 10. Infant and maternal deaths at childbirth from 1940 to 1970 (per 100,000 live births). *Source:* Office of Health Statistics Analysis, National Center for Health Statistics, Health Services and Mental Health Administration, Public Health Service, Department of Health, Education, and Welfare (1971).

transfusion technique. Prior to World War II, hemorrhage was the number one killer of women during childbirth, with 6 deaths/10,000 live births. Today, hemorrhage is still the leading cause of maternal mortality, but the incidence has been reduced to 1 death/10,000 live births.

The third advance was formal training in the specialty of Obstetrics and Gynecology. At the end of World War II (1945) less than 2,000 physicians had completed 3 years of residency training, qualifying them for specialty certification by the American Board of Obstetrics and Gynecology. Today there are approximately 20,000 physicians practicing obstetrics and gynecology in the United States and more than 10,000 are certified as specialists.

An impressive decrease in infant mortality statistics (Fig. 11) has occurred since World War II in the United States. Antibioties and improved nutrition in the first month after birth have contributed, but the major

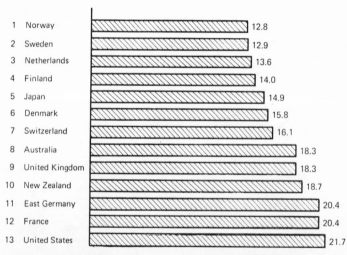

1	Norway	12.8
2	Sweden	12.9
3	Netherlands	13.6
4	Finland	14.0
5	Japan	14.9
6	Denmark	15.8
7	Switzerland	16.1
8	Australia	18.3
9	United Kingdom	18.3
10	New Zealand	18.7
11	East Germany	20.4
12	France	20.4
13	United States	21.7

FIG. 11. 1968 Infant mortality data for selected countries of the World. Rate per 1,000 live births. *Source:* Office of Health Statistics Analysis, National Center for Health Statistics, Health Services and Mental Health Administration, Public Health Service, Department of Health, Education, and Welfare (1971).

factor is the increased skill in treating the premature
infant. Three out of four deaths shortly after birth occur
in the premature. Intensive medical care for the prema- CLINICAL
ture infant has become highly sophisticated, making it MEDICINE
possible to save premature babies weighing as little as
1-2 pounds.

In 1968, the United States ranked 13th among nations in childbirth mortality (Fig. 11). The top-ranking
countries (Norway, Sweden, The Netherlands, and Fin-
land) are smaller and have more homogeneous populations. Their well-developed family planning services
prevent high risk patients from becoming pregnant. If a
woman should accidentally become pregnant, she would
be aborted. This has had an overwhelming effect on infant mortality statistics.

In the United States, on the contrary, the restrictive
atmosphere relative to therapeutic abortion (abortion
as a medical health treatment), and also relative to
family planning, has permitted high-risk pregnancies
among women with Rh sensitization, diabetes, and
toxemia. Physicians have learned to provide better care
prior to childbirth and more intensive care after childbirth. With rapidly developing contraceptive practices
and liberalized abortion and sterilization laws, United
States childbirth mortality should decrease markedly.
High infant mortality and poverty are closely associated
in the United States and infants of the economically
poorer groups suffer the higher mortality.

A major problem facing the world today is population control. Until a few years ago in the United States
contraception had such religious, social, and personal
implications that organized family planning was buf-
feted by an overwhelming undercurrent of resistance.
More recently, the cost of raising a family, overpopulation, and pollution of the environment have made the
public acutely aware of the need for family planning and
population control. Today, contraception is no longer
just a family planning problem; it is a necessity for
survival.

Methods of contraception, such as the diaphragm
and the condom, had failure (pregnancy) rates that were

unacceptable to married couples earnestly interested in practicing birth control. However, the failure rate is primarily associated with the way in which people use them rather than inefficiency of the devices. Many individuals feel these methods are inconvenient and burdensome; others use them sporadically or abandon them.

In 1962 "the pill" was found to be an effective contraceptive when taken for 20 days each month. The major drawback to the early oral contraceptive was nausea and fluid retention. Lower dose pills proved effective and caused few side effects. Associated with "the pill" is a slightly increased incidence of thromboembolic phenomena involving the clogging of blood vessels by blood clots in, for example, the legs and lungs. Minor

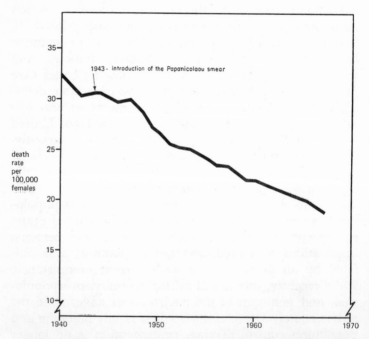

FIG. 12. Deaths from malignant neoplasms of female genital organs except breast cancers since 1940. *Source:* Office of Health Statistics Analysis, National Center for Health Statistics, Health Services and Mental Health Administration, Public Health Service, Department of Health, Education, and Welfare (1971).

and occasional side effects are fluid retention, weight
gain, breakthrough bleeding (vaginal spotting), headaches, and occasionally blood vessel spasms.
More than 8 million women take some form of "the
pill" today. It is virtually 100% effective. On another
front, the intrauterine contraceptive device (IUCD)
has become widely used as family planning centers treat
vast numbers of patients. There were 2–3 pregnancies/
100 women during the 1st year, with slightly lower rates
in subsequent years. Over 2 million women use this
method.

The average birth rate between 1953 and 1955 was
24.8/1,000 population. The effects of family planning
became evident in 1968 when the birth rate was dropped
to 17.8, the lowest in United States history.

A simple cytologic test (the Pap Test) was developed
in 1943 for detection of cancer of the cervix. Since that
time, the Papanicolaou smear has been used over 300
million times (Fig. 12). Twenty-five years ago cancer of
the genital organs, primarily cervical cancer, was the
number one killer of women. Today mortality due to
cancer of the cervix has decreased 40% due to early
detection. Annual gynecological examinations with
Papanicolaou smears ("Pap smear") have become an
increasingly accepted routine; patients on oral contraceptives or replacement estrogens must return for renewal of their prescriptions and this provides a convenient opportunity for Pap smear tests.

GERIATRICS

Geriatrics is becoming a recognized specialty in the
United States, culminating 25 years of progress in developing and implementing comprehensive programs of
care for the aged that has resulted in better understanding of the processes of aging and in the diagnosis and
treatment of the diseases of age. These advances fall into
three categories: research, clinical practice, and social
planning.

Basic research into why people age has been expanding
in the last decade. A number of theories have been ad-

vanced for aging, aging manifested as biochemical changes in tissue cells, theories based on information loss or error occurring in the sequence of genetically controlled biochemical steps by which proteins are formed.

Although advances in clinical practice in the field of aging are numerous and varied, the greatest gains are: *1) Pharmacology*. The boom in chemotherapy has perhaps helped the aged more than others. Of equal significance is the increasing sophistication of physicians in prescribing drug dosages tailored to the altered metabolism of age. This has been a slow gain, but has had a major cumulative effect on the treatment of geriatric patients. *2) Surgery*. The entire philosophy of surgery for the elderly has undergone a revolution. Not long ago if you were old you automatically were in the poor-risk category. The old person was "treated conservatively" and denied surgery even when surgery offered the best solution to his problems. Increasingly age is no barrier to surgery, provided proper precautions are taken. Today the most radical procedures are being performed where they appear to offer meaningful years of life. *3) Psychiatry*. The gains of the last 25 years have been significant. We are better able to recognize the depressions common among the elderly, and we now have drugs with which to treat them. We no longer assume senility without investigating the other causes of mental slippage, and in treating the senile we are seeking methods in the direction of better self-care. *4) Rehabilitation*. Rehabilitation of the elderly is still in its infancy in the United States. Manpower and facilities for major rehabilitative endeavors remain in short supply, and the time, effort, and cost often are staggering.

In social planning the gerontologists, a broad group that includes basic scientists, psychiatrists, public health officials, psychologists, sociologists, economists, and others, have produced a great body of knowledge that contributes to a better understanding of the attitudes, behavior, and life situations of the aged. Through the efforts of the gerontologists, clinicians have come to understand that the total patient and not a single organ or disease must be treated.

Many advances in geriatrics have resulted from the work of physicians who do not consider themselves geriatricians but who have sought ways to improve the treatment of their elderly patients. With an increasing number of elderly patients, clinicians now function as geriatricians also.

Diabetes is a good example of the gains made. It is now recognized that diabetes in the young differs considerably from the diabetes in the older patient and the treatment is vastly different. The older patient can be very well handled by the use of proper diets alone. Other "standard" methods, whether involving types of therapy, drug dosages, or test results, also have been modified by a growing awareness that what applies to the young does not necessarily apply to the old.

In the case of Parkinson's disease, the advent of cryosurgery or surgery with decreased temperature, and more recently of L-dopa, has opened interesting avenues of treatment. More cases of arthritis are benefitting from newer drugs and from the successful efforts of the orthopedic surgeons. Implanted pacemakers and artificial heart valves are keeping many elderly people from becoming cardiac cripples. The increased knowledge in the field of ophthalmology has resulted in better vision among the geriatric group. The early diagnosis of glaucoma and cataracts has resulted in a marked restoration of vision, and preventive ophthalmology is beginning to achieve sight saving in this older group.

NUTRITION

By the time of World War II an almost complete description of the nutrient requirements of man was accomplished. During the following 30 years, the fruits of this knowledge have been applied extensively. Despite the continuing existence of widespread hunger, famine, and malnutrition in the world, there are probably more well-nourished people today than ever before, particularly in the developed countries.

Sound nutrition rests on three supporting mechanisms. An inventory of foods must be available, persons must have the means to obtain, and sufficient information to make appropriate choices. The requirements of inventory, income, and information have each been augmented in the last 30 years by the work of nutritional scientists. It was fitting that the 1970 Nobel Peace Prize was awarded to a plant geneticist for efforts directed to increased food production.

The inventory of food has been increased by what is popularly called "The Green Revolution," a combination of plant breeding, plant nutrition, agronomy, and food technology directed toward more efficient crop growth, harvesting, preserving, and marketing. These activities have made the land more productive and foods more available. This revolution increased the yield of agricultural products in the United States by 40% since 1945, while cultivated acreage decreased by 11%. This revolution made the Republic of the Philippines self-sufficient in rice, its staple food, for the first time since 1903. It has increased wheat production in India and Pakistan by 30% in the last 3 years.

A similar revolution in animal science has gone on in the Western world especially where it is still feasible to feed grain to animals. The science of breeding, feeding, and marketing animals for food has greatly increased human welfare. Although it is sometimes viewed as a paradox that the nutritional requirements of food animals are better met than those of man, the fact is that the first accomplishment aids the second.

It is difficult to evaluate the role of rising economic levels in bettering nutritional status. The presence of a cash income and the availability of a variety of foods, often produced in distant areas, leads to a heterogeneous diet and away from a precarious nutritional adaptation on a monotonous and limited menu. The control of infectious disease, with the lessened prevalence of fever, has had a profound advantage for nutritional status by avoiding this metabolic deviation of nutrients to heat production in fever and to other wastage.

Yet with all the advances in scientific understanding of nutrition, public understanding of adequate nutrition has been less successful. Educational efforts of academic and public organizations and of the food industry have

been outstripped by publicized development and marketing of prepared foods. Nevertheless, there has been an
overall improvement in food choices aided by the availability of a variety of foods at practical prices. Indeed,
the variety of food available may now be to some extent
a nutritional hazard because people are tempted to
divert their money to bad choices. Recognizing this
threat, major efforts toward incorporating nutritional
education in the primary and secondary schools have
been developed recently. Another important achievement has been the development and use of reference
tables of child growth by pediatricians. The availability
of commercial feeding formulas that are nutritionally
sound, safe, and economical while perhaps less desirable
than breast feeding have made the feeding of babies a
highly successful science. The main uncertainty has to
do with goals. What, for instance, is the optimal size
and rate of gain of a child? The recognition and treatment of growth failure associated with inadequate
nutrition and of the two cataclysmic forms of this,
diseases called marasmus and kwashiorkor, are now
within the capabilities of professional persons in pediatrics. Marasmus is an extreme emaciation or wasting
away that affects young children especially. Kwashiorkor
is a very serious protein and caloric malnutrition seen
particularly in children in underdeveloped areas, almost
entirely outside this country.

The recognition that a daily intake of 1–2 mg of
fluoride will protect the teeth from caries led to proposals
for fluoridation of public water supplies. Gradually, and
with public resistance, this program has led to the
supply of fluoridated water to 80 million Americans.
This measure alone has been shown to reduce the
prevalence of dental caries by 60%.

The most prevalent nutritional disorder in the United
States is iron-deficiency anemia, especially in infants
and women of reproductive age. Where diet is only
marginally adequate, the menstrual loss of iron can have
serious consequences. Milling and sanitation procedures
have reduced the natural iron content of food. The
enrichment of certain commercial cereal foods replaces

some of the iron. The adequacy of enrichment techniques in diminishing the prevalence of iron deficiency is currently under intensive study.

Important progress in the area of vitamins has involved vitamin B_{12}, which long baffled the efforts of biochemists. In the 1950's the vitamin was isolated and characterized chemically. Vitamin B_{12} quickly became a remarkably useful medication for managing the deficiency state that characterizes pernicious anemia, a chronic progressive anemia of older adults.

Folic acid was characterized chemically in the late 1940's and it has slowly come into clinical importance as another useful vitamin employed for the management of macrocytic anemia that is associated with abnormally enlarged red blood cells. It is estimated that a third of all the patients in the Western world with this anemia are relieved by the administration of folic acid. The surge of research activity with vitamin D in the 1920's subsided for 40 years while synthetic vitamin D, calciferol, was widely used to prevent rickets in children. Rickets is a deficiency disease in which calcium is absorbed inadequately from the intestine into the blood stream. The chemical nature of the compound, 2,5-OH calciferol, was elucidated in the 1960's and shown to be effective in facilitating the intestinal absorption of calcium.

Vitamin K, discovered in research on chickens in 1935, came to have practical importance as a means of diminishing the frequency of hemorrhages at birth and consequently of birth injury to babies. An infant at birth is unable to make his own supply of vitamin K. It was important to learn that dosing a mother before delivery provided the infant with the needed vitamin and diminished bleeding from trauma during the birth.

Studies of pyridoxine, vitamin B_6, have given important insights into the mechanisms of vitamin action in the body and have also provided useful clinical procedures. Deficiencies of this vitamin can be of two kinds: acquired, or caused by a low intake of food; and genetic, the so-called pyridoxine dependency. Acquired deficiency impairs all of the several functions of pyridoxine.

It is relieved by relatively small doses of the vitamin.

Pyridoxine dependency, however, is a specific genetic defect of one or another reaction of pyridoxine; it requires very large doses of the vitamin for control. Examples of pyridoxine dependency are deficiencies of enzymes leading to epilepticlike seizures, anemia, and certain diseases involving mental retardation.

137

CLINICAL
MEDICINE

Intravenous feeding as an emergency procedure when oral intake is impaired has been a notable advance in patient care resulting from the science of nutrition.

Alcoholism, affecting at least 9 million people in the United States, is an important cause of malnutrition. An earlier assumption that alcohol acts solely by replacing food intake has not been sustained. Alcohol is a cell toxin and above certain limits, perhaps in the order of 20–30 % of the daily calories, will damage organs despite a good diet. The problem of cirrhosis of the liver, a common consequence of excessive alcohol intake and the fifth most common cause of death in the middle-aged American, will require more than a good diet for its solution.

Food enrichment with selected nutrients such as thiamin, retinol, calciferol, riboflavin, niacin, and iron originated with the Food and Nutrition Board of The National Research Council in 1943. The concept that nutrients lost in food processing should be replaced is feasible, economical, and efficient. Diseases caused by thiamin, riboflavin, and niacin deficiencies are rarely seen now. A dramatic demonstration of the effectiveness of the enrichment of rice with thiamin showed that beriberi could be virtually eradicated in the rice-eating Philippines.

Through the awareness that pernicious anemia results from failure to absorb vitamin B_{12}, there is an effective treatment for this formerly progressive and fatal disease. The neurologic complications of pernicious anemia, once seen frequently, are now rare. An important group of the polyneuritis diseases including those associated with chronic alcoholism caused by a dietary deficiency can be treated and reversed by the administration of thiamin.

CHAPTER III   Neurology, a branch of clinical medicine that deals with the diagnosis and treatment of diseases of the nervous system, was once closely allied to psychiatry.

Electrical changes in the brain in man were first demonstrated in 1929 and it was shown that brain waves could be recorded through the intact skull. The term electroencephalogram was coined for such a brain wave record obtained with an instrument called an encephalograph. When it was shown that characteristics of brain waves could be used as indices of brain disease, a new approach to studying brain mechanisms and brain disease developed. Clinical electroencephalography, beginning about 1935, found major use in the diagnosis and delineation of various types of epilepsy. Another technique, called electrocorticography, in which electrodes are placed directly into the tissue of the cortex or outer part of the brain itself, provides a more precise diagnosis and localization of brain activity.

Although brain wave measurement contributed immeasurably to the study of epilepsy, a significant advance came with the development of effective drugs. Diphenylhydantoin, a drug used widely in the treatment of the major epileptic convulsions, was introduced in 1938. Other drugs of this type, as well as barbiturates and other anticonvulsive compounds, have since been introduced for the treatment of major seizures and those which are less severe.

Another diagnostic procedure developed during the last 30 years is X-ray analysis of brain blood vessels following the injection of radioactive material. This technique, referred to as cerebral angiography since the "cerebrum" is the principal and surface part of the brain, came into widespread use about 1940. It is especially valuable in the diagnosis of vascular diseases such as abnormal narrowing or blockage of major blood vessels to the brain or an abnormal widening and weakening of the vessel walls. Cerebral angiography is also helpful in the diagnosis of localized abnormalities in the brain such as tumors or tumorlike growths and blood clots under the outer membrane covering the brain.

Scanning of the brain following the injection of radioactive isotopes, in general use for only 15 years, is of
outstanding value in the diagnosis not only of tumorous
growths in the brain but also in the diagnosis and interpretation of other brain lesions, including brain abscesses,
blood clots, blockage of blood vessels, and malformations
of arteries and veins.

Radioactive materials injected directly into the spinal
canal are used in the diagnosis of hydrocephalus ("water
on the brain," an accumulation of fluid in the brain
cavities that causes the brain to thin and the head to
enlarge) and other abnormalities in the circulation of
fluid other than blood in and around the brain and
spinal cord.

An infectious disease, central nervous system syphilis,
which was once the most common of the organic diseases
of the nervous system, has been eradicated almost completely by prompt and early treatment of syphilis with
penicillin and related drugs. Antibiotics have also
significantly decreased the mortality rate and effects of
the acute bacterial meningitis diseases. Streptomycin,
either alone or with other drugs, has decreased the
mortality and morbidity from tuberculous meningitis.
Viral diseases of the nervous system are still enigmas as
far as treatment is concerned. One of the most significant
advances in medicine in recent years has been the development of a vaccine to prevent poliomyelitis, a viral
infection involving inflammation of the spinal cord
causing infantile paralysis. This disease is now almost
nonexistent (Table 1).

However, there are some diseases, as yet unconquered,
that are caused by the so-called "slow viruses." Certain
neurologic disorders of obscure origin are now known
to result from a prior infection by a slowly growing and
late maturing virus. One such disorder is kuru, a disease
found in the mountains of New Guinea. Kuru occurs in
young and adolescent children, causing rapid incapacita-
tion. The disease produces degeneration of the brain and
spinal cord and death occurs at an early age.

Although marked advances have been made in the

diagnosis and treatment of one of the commonest of the metabolic disorders, diabetes mellitus, neurologic complications (both peripheral nerve and cerebrovascular) still occur frequently. Other metabolic disorders, however, some only recently described and delineated, can be avoided or controlled to some extent by dietary management.

A form of paralysis and mental disorder, once thought to be a degenerative disease, is now known to be caused by a genetically determined disturbance of copper metabolism. The disease can be treated effectively by drugs that decrease the concentrations of copper in body tissues.

The vascular diseases affecting the nervous system constitute an extremely important group of neurologic conditions. With the advent of techniques for taking X-ray pictures of blood vessels, it has been found that, in many patients in whom clogging of arteries within the brain had been suspected, the blockage is actually in one of the blood vessels in another part of the body. In selected cases, surgery has been helpful. Anticoagulants also have been used for the treatment of blood clotting complications and so-called "transient ischemic attacks," temporary conditions of inadequate blood flow in the brain.

Techniques for recording electric currents in muscle contractions have aided in the evaluation of diseases of muscle. The muscular dystrophies, inborn abnormalities of muscle, diseases largely of childhood, have recently been more completely classified and differentiated. The presence of specific enzymatic changes in the blood has been found to aid in diagnosis.

Parkinson's disease is known to be a metabolic disturbance involving a deficiency of dopamine and other related amines in certain parts of the brain. The chemical, L-dopa, is much more effective than any previous drug because it decreases the rigidity and slowness of movement as well as the tremor. This treatment, still in the early stages, has promise as one of the most significant advances in therapy of this crippling disease.

Probably the most important advances in psychiatric care in the past 30 years, in the order of their appearance, have been: *1*) the advent of various body (somatic) treatments, one of which, electric shock, is still in vogue and is almost specific for some forms of depression; and *2*) lessons learned from World War II: *a*) that every man has a psychological breaking point and that careful attention to morale factors can postpone or aid in preventing the break; and *b*) that early treatment in familiar surroundings can prevent the emotional illness from becoming chronic. This knowledge has lessened the number of psychiatric casualties in warfare and formed the basis for psychiatric intervention in the crises of civilian life.

CLINICAL MEDICINE

The contributions of the biological sciences led to the introduction of useful drugs in the treatment of the mentally ill. Mental illness is a major problem. Most hospital beds in the United States are occupied by mentally ill patients, and 1 million are treated in public and private hospitals at a cost of over a billion dollars each year. The indirect cost of mental illness, including the economic loss of productivity and income, is in excess of 3 billion dollars yearly. Within recent years, the introduction of such drugs as chlorpromazine and similar psychotropic drugs has contributed to the decline in the number of patients occupying beds in psychiatric hospitals (Fig. 13). This is in spite of an increasing population and an anticipated increase in the number of psychiatric patients. Today more and more mental patients are treated in clinics with drugs and are not hospitalized. The patient avoids the stigma of being admitted to a mental hospital, remains with his family, and carries on a productive occupation.

The new antidepressant and antianxiety drugs, while not living up to advance billing, have enabled psychiatrists to treat patients at work and at home. An outstanding example is lithium carbonate, useful in preventing the disheartening appearance of mania in patients prone to manic depressive psychosis.

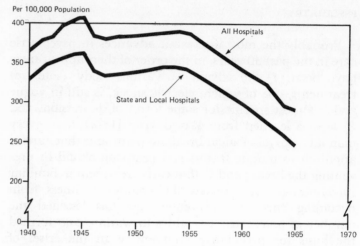

FIG. 13. Resident patients in hospitals for psychiatric care. *Source: Health, Education, and Welfare Trends,* 1966–67 edition, Department of Health, Education, and Welfare, p S-38 (1968).

OPHTHALMOLOGY

As in other areas of medicine, the antibiotics, corticosteroids, and antimetabolites have had a major impact in the treatment of infections, inflammations, and certain tumorlike growths of the eye. Improvements have been made in the treatment of growths on the retina by a combination of antimetabolites, radiation, and coagulation of tissue by an intense beam of light (photocoagulation). The eye and its associated parts become involved in a number of systemic diseases, including neurological diseases, and a new subspecialty, neurophthalmology, has developed as an intermediate between the areas of ophthalmology, neurology, internal medicine, and pediatrics. Because of its accessibility, the retina of the eye (which itself is an extension of the brain) is an especially favorable site for neurophysiological studies. The recent techniques of visualization of the blood vessels of the eye, using fluorescent dyes, and of photocoagulation produced either with the laser beam or with xenon light, have led to important advances in the diagnosis

and treatment of lesions in the blood vessels of the retina.
For the first time certain retinal lesions can be properly
diagnosed and treated, and abnormalities of the retinal
blood vessels associated with diabetes, sickle cell anemia,
blood vessel occlusion, and other vascular abnormalities
can now be successfully recognized and treated.

Retinal degeneration and blindness in premature
infants is now avoided through a proper use of oxygen
therapy; blindness from this cause has been greatly
reduced. The vascular abnormalities associated with
diabetes are especially prominent in the retina.

Advances in examination techniques and surgical
methods in connection with retinal detachment and
holes in the retina have led to a greatly improved future
for patients with these defects.

The cornea is the transparent membrane over the
pupil of the eye through which light goes to the retina
within the eyeball. Because the cornea is simple in structure and contains no blood vessels, it is particularly well
suited for studies on how chemicals get through membranes and for studies of tissue reactions in transplantations. It has been known for many years that the cornea
could be transplanted successfully from one person to
another when the transplantation of other tissues and
organs failed. An explanation has been provided by
studies done in the early 1950's, improvements have
been made in the surgical techniques of corneal grafting,
and flush-fitting corneal contact lenses have been developed.

In galactosemia, a disease caused by an inborn error
in the metabolism of galactose, large quantities of sugar
alcohols accumulate in the lens of the eye causing cataract. Similar changes have been found recently in the
lenses of diabetic patients with cataracts. Another recent
finding is that cataracts are among the defects in infants
resulting from German measles in the mother. Cataract
surgery has been improved technically to such an extent
that the procedure in skillful hands is now about 98 %
successful.

In glaucoma the flow system of the eyeball is defective
so that pressure within the eyeball becomes abnormally

high, causing degeneration of the retina and blindness. It has been found that in adults glaucoma is either of an open-angle or of a closed-angle type ("angle" referring to the angle in the eye through which the fluid flows out). Open-angle glaucoma is known to be an inherited disorder; in most cases, medical treatment is successful and blindness can be prevented. A major finding in 1958 was the observation that certain enzyme inhibitors act as secretory inhibitors and reduce intraocular pressure. Other studies have focused on the flow and pressure of the internal liquid of the eye. Techniques are now available for accurately measuring this intraocular pressure; these provide a diagnostic aid in early identification of glaucoma.

One of the most puzzling problems and difficult conditions to treat in all ophthalmology is a group of inflammations that affect the different parts inside the eyeball. One of these inflammatory conditions has been found to be associated with an infectious agent in the retina and another with a parasitic worm in the vitreous humor, retina, and choroid coat of the eye.

In the past 30 years there have been important improvements in ophthalmological instrumentation, particularly in microsurgery. Developments of microsurgical tools have been stimulated by the increased use of the operating microscope in ocular surgery.

OTORHINOLARYNGOLOGY

New instruments, such as the operating microscope and the endoscope—used in direct examinations of certain body cavities and passageways—have altered profoundly both the medical and surgical treatment of diseases of the ears, nose, and throat. An underlying reason for the advances in this field of medicine known as otorhinolaryngology was the development of antibiotics that changed the focus of surgical procedures from the evacuation of pus toward reconstructive surgery. The complications of ear infection (otitis media) have been reduced more than tenfold. Extensive tumor surgery of the head and neck and reconstruction by use of skin flaps and preservation of the voice box (larynx)

in selected cases of throat cancer by a particular surgical
procedure (supraglottic laryngectomy) have become a
reality in the past 30 years.

The modern era of surgery for abnormal growths of
nerve tissue associated with the hearing mechanism
(acoustic neuroma) dates to 1961. With the advent of
more sophisticated clinical tests and hearing tests, it
became possible for the medical specialist to diagnose
acoustic tumors early. Armed with the operating microscope, he had new weapons with which to deal with small
acoustic tumors. This reduced the morbidity and mortality of the primary neurosurgical approach.

With the advent of antibiotics, the simple mastoid
operation for acute pus-forming inflammation of the ear
and mastoid has become rare. However, these destroyers
of the middle ear—chronic inflammation and mastoidi-
tis—have not abated. Operations to eliminate infections
and reconstruct the hearing of the middle ear have been
perfected since 1954.

Many methods of reconstructing the damaged ear
drum and bones of the middle ear have been devised.
They all involve microsurgical techniques to rearrange
or replace portions of the chain of ear bones (ossicles)
and graft perforations of the eardrum (tympanic membrane). The most commonly used tissues for repair of the
eardrum today are the tissue sheet enclosing a muscle, or`
the skin, or both. Recently, homografts of the eardrum
and attached bones have been used in an attempt to
restore the hearing mechanism.

Probably one of the most publicized and most gratifying operations of modern otology is the surgical removal
of one of the ear bones, an ossicle called the "stapes"
(stapedectomy or oval window fenestration). This bone
is called "stapes" because its shape is something like
that of a horse stirrup. The stapedectomy operation is
ordinarily done for otosclerosis in which the stapes is
abnormally hardened and fixed by new bone growth,
but it can also be done to relieve the abnormal rigidity
of the stapes caused by sclerosis or abnormal hardening
of the eardrum or by congenital fixation. The incidence

of otosclerosis in the general population is about 1 person in every 100; however, only 1 out of every 1,000 of these will have clinical hearing loss. Otosclerosis is unequally distributed between the sexes, being present in 1 of every 8 females but in only 1 of every 15 males. Surgery for otosclerosis has gone through many stages, from a one-stage fenestration (a surgically produced opening to improve hearing) of the 1940's, through a stapes mobilization procedure in 1952, to the stapedectomy, first done in 1957. The operation generally consists of removing the fixed stapes and replacing it with an artificial part (a prosthesis). The results of this operation are very good and the great majority of patients obtained improved hearing.

With all the advances in surgical techniques and equipment, chances of restoring functional hearing in patients with congenital malformations of the ear have increased remarkably. Since the inner ear has a different embryological origin than the middle and external ear, the inner ear usually functions normally in patients with congenital deformity of the external canal and middle ear.

The accurate measurement of hearing and hearing loss is vital to the assessment of communication handicap. The most common form by which the results of hearing tests are displayed is called an audiogram. The concept of speech audiometry is probably the most important single contribution to the evaluation of hearing impairment. It provides indispensable information about the extent of the patient's communication handicap, helpful in hearing aid evaluations, and is important in differentiating the location of the hearing disorder. But there has been little progress in treatment of sensory hearing loss, deficits associated with failure of the inner ear to function properly. Research has been directed toward understanding the process whereby certain important cells of the ear, called hair cells, are damaged by industrial noise, gunfire, toxic drugs such as kanamycin and ethacrynic acid, and progressive inherited losses. A reasonably good idea has been developed of the hazard associated with specific noise or drug exposures. However, attempts to treat such sensory–neural damage by blocking certain groups of nerve cells or ganglia (stellate block) or by administration of various substances, such as nicotinic acid, vitamin A, or adenosine triphosphate, have failed to produce improvement in hearing. Unfortunately, hair cells do not regenerate.

Even though the most dramatic changes have occurred in the field of otology, many important contributions have also been made in the field of head and neck cancer surgery and nasal physiology and also by the technique of endoscopy, the examination of cavities such as those of the nose, ear, and throat with an instrument called an endoscope.

Cancer of the oral cavity, pharynx and larynx has been increasing steadily in the past 20 years. The otolaryngologist is thus faced not only with detection and treatment but reconstruction as well. Surgery for cancer of the head and neck ideally should embody two main concepts: removal of the primary tumor, and total reconstruction and rehabilitation, both physiologically and cosmetically. Conservation surgery of the larynx means adequate removal of the tumor while preserving the functions of the larynx. Fifty percent of all laryngeal and laryngopharyngeal cancers are amenable to operations that preserve the voice function.

Recently the use of the operating microscope has been used in the removal of laryngeal lesions, such as small nodules, polyps, and horny lesions, greatly increasing the accuracy of these operations and enabling a higher percentage of voice rehabilitation.

In the past 30 years, there has been renewed interest in nasal physiology. The nose serves the body for smell (olfaction), humidification and cleaning the air we breathe, and proper respiratory resistance. Prolonged increased nasal resistance due to nasal obstruction can cause pulmonary disease and secondary heart disease. Studies of airflow through the nose in normal and pathological states have contributed to an intelligent approach to the correction of internal and external nasal deformities.

Due to the increased use of automobiles and snowmobiles, accidents and fractures of the facial bones have increased greatly. Repair of both these injuries has undergone extensive improvement since World War II.

RADIOLOGY

Rapid advances have been made in diagnostic radiology in the past 30 years resulting from improvements in equipment, technique, methods of examination, and radiological research.

In the early 1940's automatic phototimers came into use for controlling predetermined amounts of radiation in the taking of X-ray pictures. Later the brightness of the fluoroscopic image was greatly increased by image intensifiers, leading to better picture taking, seeing, and recording. X-Ray studies have been improved, some being able to focus on spots as small as a third of a millimeter (about one-hundredth of an inch), and a tube has been developed that, together with the use of high voltage, permits very brief exposures. A method called tomography has been developed for getting clearer pictures by blurring out those parts not wanted in the X-ray picture, a method very useful in the middle ear where minute detail must be seen. Photographic methods have been developed that permit increased speed of film exposure, reduce undesirable exposure of the patient, and increase information obtainable from the picture. Improvement in X-ray pictures of soft tissues makes possible diagnosis of about 80% of breast cancers. Magnification techniques make visible X-ray pictures of blood vessels in the brain, kidney, and heart, and of the chest in children.

Methods of examination by radiologic techniques have been improved and widely used in connection with the brain, spinal cord, aorta, large arteries, bile ducts, body joints, gastrointestinal tract, urinary tract, and lymph glands. Improvements have resulted partly from the use of better contrast agents injected into blood vessels and other cavities. Extensive research has led to these improvements, and special training has provided skilled personnel. Major advances in 30 years have been made

also in the equipment that produces radiation treatment.
The first betatron was developed in 1940, linear accelerators became practical after 1948, and the first cobalt 60 unit came into use in 1951. These penetrating radiations now achieve adequate radiation dosages to deep seated tumors, less reaction to the skin of the patient, less damage to bones, and with improved results. Comparable improvements have been made in radium therapy. Research on the effectiveness of radiations of different quality have led to improvements of dose–time relationships in treatment. Future advances are expected from studies of the clinical use of certain chemical reagents in conjunction with radiation treatment. Although the use of radiation treatment after surgical operation has had full acceptance, and still is the most frequently used way of combining radiation and surgical treatments, in recent years preoperative radiation therapy has been found advantageous particularly in the treatment of certain malignant tumors of the esophagus, lung, kidney, bladder, and the lining of the uterus.

 CLINICAL
MEDICINE

An important recent development is the increasing evidence that long-term control of Hodgkin's disease and related disorders can be achieved by applying high doses of radiation to certain large areas of the body. Hodgkin's disease is a chronic disorder of the lymph glands that may be associated with disturbances of the spleen, liver, kidneys, and blood vessels. The treatment has increased the survival of these patients. During the past several years, computers have come into use in the determination of radiation dosages in the treatment of patients. Through the development of picture transmission over long distance telephone, a treatment center can provide consultative advice, treatment planning, and computer-determined radiation dosages to distant and smaller treatment installations. There is a great need for additional trained radiotherapists and paramedical personnel in this area.

NUCLEAR MEDICINE

Nuclear medicine, which involves the application to medicine of nuclear reactions and the use in medicine of chemical compounds labeled with nuclides (radioactive atoms or parts of atoms), has developed since World War II. The appearance of the cyclotron was followed by the production of many pharmacologically important radioelements, such as sodium, phosphorus, iron, and even carbon, and was about to be followed by the birth of nuclear medicine in 1939, when World War II shrouded all cyclotrons in secrecy. During the war years a few unclassified biological studies were made, including a surgical study with tagged serum albumin, the use of radioiodine in the treatment of thyroid cancer, and the use of tagged compounds (the "atomic cocktail") in the treatment of leukemia and polycythemia (increased red blood cells in the blood). But, for nuclear medicine, the most important event in the war years was the advent of the nuclear reactor, which could produce a greater volume of medically useful nuclides than could the cyclotron. About 1950, the pattern for nuclear medicine was in part set by five events: *a*) the rapid realization that the use of radioiodine was likely to be more successful for benign hyperthyroidism than for thyroid cancer; *b*) the use of serum albumin labeled with radioiodine to measure plasma volume, one of several things that led to the development of radioactivity instrumentation by industry; *c*) the discovery that the blood plasma and the red blood cells could be labeled with different radioactive tags and, since red cells and plasma could be separated, routine clinical tests could be made for red cell and plasma volumes, spaces, and clearances; *d*) the discovery that vitamin B_{12} could be labeled with radioactive cobalt which led to a routine clinical test for pernicious anemia; and *e*) the discovery that certain radioisotopes would concentrate in tumors which, when near the skin, could be roughly localized by instruments detecting their radiations. Also that tumors of the brain could be localized, as could the size and shape of the thyroid gland after its exposure to radioiodine. All of this led to sophisticated techniques for mechanically scanning suspect areas of the body, techniques that were introduced into nuclear medicine in 1950. Other studies were done with radionuclides

also which did not result in specific nuclear medicine
tests.

With the development of the atomic bomb and atomic
bomb explosions, studies were made on radioactive
fallout, with respect to such problems as components of
the fallout possibly causing leukemia and cancer and
the possibility of injury from fallout radiation. In the
1960's short-lived radioisotopes were introduced and
some of them used successfully for test or diagnostic
purposes in nuclear medicine, for example, in brain
scanning.

Scientific advances were made, such as immunologic
assays, and labeling and measurements with tritium.
Thus, the basic nuclear medicine of the 1950's was
modified and further developed in the 1960's. And this
has been accompanied by developments in biomedical
engineering.

BIOENGINEERING

The greatest development of biomedical engineering,
sometimes called bioengineering or medical engineering,
has taken place during the last 2 decades when attempts
have been made to define a separate interdiscipline in
which the tools of engineering are to be applied to
medicine to improve health care. Examples of such
applications would be in the areas of artificial limbs or
organs, screening and diagnostic procedures, operating
room and recovery room instrumentation, and patient
record systems. But effective communication between
physicians and engineers has been limited, especially
by the fact that each discipline has its own unique
vocabulary and educational programs. Although training
programs have attempted to bridge this communications
gap, the number of engineers and physicians exposed
to each other's discipline has been relatively small.
Furthermore the disciplinary practices differ. Engineering deals largely with synthesis, such as the design of new
electronic devices. Medicine deals largely with analysis,
such as physical diagnosis; and living systems, such as
those of the body, pose some very difficult problems.

Routine use of the artificial kidney had to be postponed until the late 1950's when relatively permanent access to blood circulation was made possible by the development of tubes made of Teflon and silicon rubber components. The need to diffuse a patient's blood across a semipermeable membrane called for a particular internal design of the artificial kidney. These and other bioengineering problems had to be solved.

The most intensive interdisciplinary activity has been directed to the problem of developing an implantable artificial heart, the need for which is evident from the death rates from coronary heart disease (heart attacks.) Although complete success in this has not yet been attained, even in animal experimentation, improvements in materials, construction techniques, and increasing control of variables indicate that progress will be forthcoming in the next decade leading to extended periods of survival in animals in which artificial hearts have been implanted, which means progress toward the development of a usable device for patients.

Advancements in biomedical electronics were facilitated in the 1940's and 1950's with the introduction of the transistor and, more recently, of small solid-state devices and other integrated circuit devices, the initial applications of which were commercial or military rather than medical. Recently a number of transistorized electrical stimulators have been put to medical use. Since 1960 thousands of cardiac pacemakers have been implanted in patients to help treat blockage of the heart beat; mortality has been low, but limited battery life and fragility of electrodes still are bioengineering problems that are being solved by application of new knowledge. Pacemakers run by atomic energy have been developed that may prove to be more reliable than battery-operated types. These are in the experimental stage at the present time.

Electronic stimulation of certain vascular nerves has had limited success in controlling hypertension and has been used intermittently for relief during episodes of angina pectoris, severe constricting chest pains due to disturbance of coronary vessels in the muscle of the heart.

CLINICAL
MEDICINE

Military research on medical instrumentation and data management also has led to the development and use of a number of physiological sensors, medical acquisition devices, special purpose computers, and medical software for mobilizing body systems in health and disease. Examples include electrodes for long-term sensing of the electrical activity of the heart; devices for continuous measurement of gases in expired air, for accurate automatic measurement of blood pressure, for determining color vision thresholds and for testing stereoscopic vision; simple special purpose computers for analysis of heart rhythm and analysis of the electrical activity of the brain; and telemetry systems to receive a variety of physiological signals from persons at a distance during normal activity and during exposure to laboratory and operational stress. These developments have been accomplished with bioengineering participation. Many of these and other medical instrumentation developments have been incorporated into research laboratories and medical and surgical treatment facilities.

The variety of electronic instruments that have been added recently to operating rooms, recovery rooms, and intensive care areas have been to monitor continuously as many important variables in the patients as possible, thereby reducing complications and mortality. Changes in the patient's condition are electronically and automatically detected and the attention of nurses and doctors attracted by an appropriate signal. Indications are that manufacturers are competing to develop more elaborate and more effective monitoring systems.

These electronic instrumentation developments have focused attention on safety standards and procedures in hospitals, and interdisciplinary programs have been initiated to aid in the design of safe medical instruments and to recommend procedures to medical personnel that will minimize patient risk. Military studies on non-inflammable and nontoxic aircraft materials and on improved fire-extinguishing materials also have civilian hospital application.

CHAPTER III   The past decade has been called the age of the computer. The computer has come into more general use to handle the hospital and medical center problem of medical information storage and retrieval, a problem of increasing magnitude as medical care expands. The potential of computers has led computer manufacturers to support the development of medical information and retrieval systems.

One aspect of the total hospital information system is the automated clinical laboratory. More critical diagnostic techniques have required better laboratory analyses of samples from larger and larger numbers of patients. Data from the laboratory AutoAnalyzer can be fed either into a large computer center or into a minicomputer, depending on the needs and orientation of the hospital. It is possible that several minicomputers hooked into a computer center will provide the ultimate solution to the problem of automating certain aspects of patient care, including the multiphasic screening of large numbers of patients.

REHABILITATION

As a result of the increasing control of communicable and infectious diseases, of the greater availability of medical and hospital care, of better nutrition, of expanded education, of better housing, and an unprecedented standard of living in the developed countries, hundreds of thousands of persons are alive today who would have died earlier in this century with the same medical problems. Yet many of those who survived have to live with a residual disability.

Originally concerned mainly with orthopedic and traumatic disability, rehabilitative medicine found that cardiac and respiratory cripples far outnumber those with broken bones. Rehabilitative medicine plays its role in that period between the bed and the job when the fever has gone and the stitches are out. But it is not a passive convalescence medicine; it is a dynamically active medicine which applies the integrated skills of the rehabilitation team: the physician, physical therapist, occupational therapist, nurse, social worker, counsellor, and others. They seek to eliminate or minimize physical disabilities and to retrain the patient to live effectively with the abilities he has left.

Although not a rehabilitative measure, special means of getting the seriously ill or injured person to a hospital is of the greatest importance to early care and survival, whether with or without a residual disability. One of the most important achievements has been the military technique of aeromedical airlift which is now being adopted by civilian communities. More use of this method of rapid transportation should come in the future. Meanwhile, sophisticated equipment for care of patients in aircraft and a special medical airlift twin-engine jet, the "Nightingale," have been designed by the combined efforts of aeronautical engineers and Air Force medical personnel. These accomplishments should facilitate the future development of civilian airlift systems.

CONCLUSION

The accomplishments of the past 30 years, many of which were not dreamed of even 30 years ago, constitute a most important segment of the total improvements that have been made in the medicine of our Western culture during the last 300–400 years. Unlike the doctor of the 16th–17th centuries, the physician today does not rely on bleedings or purgings or the administration of herbal concoctions prescribed according to the positions of the planets to readjust imagined humors of the body. Yet the present and that remote time are historical periods that have certain similarities; both are periods of great voyages of discovery, then over the two-dimensional ocean surfaces, now through three-dimensional outer space. Both are periods of serious social, economic, and ideational upheavals associated with profound changes in human values and standards, changes of great concern and fear to the peoples of the two periods, and both periods are associated with scientific activities, primitive protoscience then, a highly sophisticated science now. In the 16th–17th centuries, of much concern to the scholastics who evidently conceived of all truth as being contained in the writings of Aristotle and the Bible, were the alchemists, who in their search for the elixir of immortality and for methods by which to transmit base metals into gold were actually beginning to seek knowledge from nature through experimental procedures.

These procedures continued to evolve and modern experimental science gradually arose in the form of chemistry and the other areas of natural philosophy, which, eventually, in the field of medicine, took the highly developed forms of modern biology and the many other interlocking specialties of basic and clinical medical science. The modern physicians, greatgrandchildren of the ancient alchemists, no longer seek for an elixir of immortality but they do continue to make advances toward a longer and healthier life for their patients and, through preventive measures, for those not yet their patients.

The complicated and often expensive discoveries and advancements have been achieved at a cost that fades into insignificance. Consider again the infectious disease, tuberculosis, a disease that in the past (1940) has killed as many as 46 persons/100,000 population per year. Today in this country with its 200,000,000 population, this would amount to 92,000 people/year. From 1940 to 1965, the average life expectancy of a wage earner in this country increased from 62.9 to 70.2 years.

The investment in research and medical care (Fig. 14) is minor when compared to the economic benefit from gainful employment of experienced people whose life is extended at least 7.3 years. Using the 1959 median income figure of $5,660.00, then these 92,000 persons earned over 38 billion dollars over a 7.3-year period. Even if we assume that the cost of developing, producing, and distributing streptomycin and other antibiotics over that 10-year period is $20 million, the cost of the development and use of the drug is insignificantly small as compared to the dollar gain. In many similar ways, medical sciences have provided immeasurable benefit to

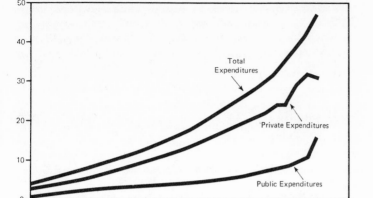

Billions of Dollars

FIG. 14. Private and public expenditures for health and medical care. *Source: Health, Education, and Welfare Trends,* 1966–67 edition, Department of Health, Education, and Welfare, p S-44 (1968).

human welfare. The expense in time, effort, and dollars may seem initially high, but the investment pays great dividends in economic and social terms. Progress has been phenomenal and the promise for further advancements and discoveries is the hope of all mankind.

SELECTED ADDITIONAL READING

Parasitic Diseases

1. CLARK, W. W., AND B. MACMAHON (editors). *Preventive Medicine.* Boston: Little, Brown, 1967.
2. COATNEY, G. R. Pitfalls in a discovery: The chronicle of chloroquine. *Am. J. Trop. Med. Hyg.* 12: 121–128, 1963.
3. DUBOS, R. J., AND J. HIRSCH (editors). *Bacterial and Mycotic Infection of Man* (4th ed.). Philadelphia: Lippincott, 1965.
4. FOX, J. P., C. E. HALL AND L. R. ELVEBACK. *Epidemiology: Man and Disease.* New York: Macmillan, 1970.
5. KILBOURNE, E. D., AND W. G. SMILLIE (editors). *Human Ecology and Public Health* (4th ed.). New York: Macmillan, 1969.

Drugs

6. CHAIN, E. B. Academic and industrial contributions to drug research. *Nature* 200: 441–451, 1963.

7. GAFFNEY, T. F., L. T. SIGELL, S. MOHAMMED AND A. J. ATKINSON. Clinical pharmacology of antihypertensive drugs. *Progr. Cardiovascular Diseases* 12: 52, 1969.

8. GOODMAN, L., AND A. GILMAN. *The Pharmacological Basis of Therapeutics* (4th ed.). New York: Macmillan, 1970.

9. RATCLIFF, J. D. *Yellow Magic, The Story of Penicillin.* New York: Random, 1945.

Immunology

10. CALNE, R. Y. *Gift of Life, Transplantation.* Aylesbury, Bucks, England: Medical and Technical Publ. Co. Ltd., 1970.

11. EDELMAN, G. M. The structure and function of antibodies. *Sci. Am.* 223 (2): 34, 1970.

12. LAWRENCE, H. S., AND M. LANDY (editors). *Mediators of Cellular Immunity.* New York: Academic, 1969.

13. RAPAPORT, F. T., AND J. DAUSSET (editors). *Human Transplantation.* New York: Grune & Stratton, 1968.

Cardiovascular Surgery

14. BARNARD, C. N. Human cardiac transplantation: An evaluation of the first two operations performed at the Groote Schuur Hospital, Cape Town. *Am. J. Cardiol.* 22: 584–596, 1968.

15. COOLEY, D. A., AND M. E. DEBAKEY. Surgical considerations of excisional therapy for aortic aneurysms. *Surgery* 34: 1005–1020, 1953.

16. DEBAKEY, M. E. Basic concepts of therapy in arterial disease. *J. Am. Med. Assoc.* 186: 484–498, 1963.

17. HARKEN, D. E., L. B. ELLIS, P. F. WARE AND L. R. NORMAN. The surgical treatment of mitral stenosis. *New Engl. J. Med.* 239: 801–809, 1948.

Cardiovascular Disease

18. BURCH, G. E., AND N. P. DEPASQUALE. *Hot Climates, Man and His Heart.* Springfield, Ill.: Thomas, 1962.

19. DAY, H. W. An intensive coronary care area. *Diseases Chest* 44: 423, 1963.

20. FRIEDBERG, C. K. *Diseases of the Heart* (3rd ed.). Philadelphia: Saunders, 1966.

21. GOULD, S. E. *Pathology of the Heart and Blood Vessels* (3rd ed.). Springfield, Ill.: Thomas, 1968.

22. HURST, J. W., AND R. B. LOGUE. *The Heart* (2nd ed.). New York: McGraw-Hill, 1970.

Renal Disease

23. DIXON, F. J. The role of antigen-antibody complexes in disease. *Harvey Lectures* 58: 21, 1962.

24. MERRILL, J. P., J. E. MURRAY, J. H. HARRISON, E. A. FRIEDMAN, J. B. DEALY, JR. AND G. J. DAMMIN. Successful homotransplantation of the kidney between non-identical twins. *New Engl. J. Med.* 262: 1251, 1960.

25. MERRILL, J. P., G. W. THORN, C. W. WALTER, E. J. CALLA-
HAN III AND L. H. SMITH. The use of an artificial kidney. I.
Technique. *J. Clin. Invest.* 29: 412, 1950.

Pulmonary Diseases

26. LIEBOW, A. *The Lung.* Baltimore: Williams & Wilkins, 1968.
27. WAKSMAN, S. A. *Streptomycin.* Baltimore: Williams & Wilkins, 1949.

Hematology

28. CASTLE, W. B. Progress in hematology, 1950–1960. In: *Year Book of General Medicine,* 1961–1962.
29. KARNOVSKY, M. L. Metabolic basis of phagocytic activity. *Physiol. Rev.* 42: 143–168, 1962.

Endocrinology

30. ASTWOOD, E. B., AND C. E. CASSIGY (editors). *Clinical Endocrinology.* New York: Grune & Stratton, 1968.
31. BONDY, P. K. *Duncan's Diseases of Metabolism.* Philadelphia: Saunders, 1969.
32. WILKINS, L., R. M. BLIZZARD AND C. J. MIGEON (editors). *The Diagnosis and Treatment of Endocrine Disorders in Childhood and Adolescence.* Springfield, Ill.: Thomas, 1965.

Medical Genetics

33. McKUSICKS, V. A. *Human Genetics* (2nd ed.). Englewood Cliffs, N. J.: Prentice-Hall, 1969.

Obstetrics and Gynecology

34. COWDRY, E. V. *The Care of the Geriatric Patient.* St. Louis: Mosby, 1963.
35. HOFFMAN, A. M. *The Daily Needs and Interests of Older People.* Springfield, Ill.: Thomas, 1970.
36. PAPANICOLAOU, G. N., AND H. F. TRAUT. *Diagnosis of Uterine Cancer by the Vaginal Smear.* New York: Commonwealth Fund, 1943.
37. TIETZE, C. The current status of contraceptive practice in the United States. *Proc. Rudolf Virchow Med. Soc. City N. Y.* 19, 1960.
38. TIETZE, C. Oral and intrauterine contraception: effectiveness and safety. *Intern. J. Fertility* 13: 4, 1968.

Nutrition

39. CORNFIELD, J., AND S. MITCHELL. Selected risk factors in coronary disease. *Arch. Environ. Health.* 19: 382, 1969.
40. ROBERTS, L. M. World prospects for increasing foods of plant origins. *J. Am. Vet. Med. Assoc.* 153: 1840, 1968.

41. WOHL, M. G., AND R. S. GOODHART (editors). *Modern Nutrition in Health and Disease. Dietotherapy* (4th ed.). Philadelphia: Lea & Febiger, 1968.

Neurology

42. BRAZIER, M. A. B. *A History of the Electrical Activity of the Brain: The First Half Century.* New York: Macmillan, 1961.
43. MOORE, J. B., JR. The epidemiology of syphilis. *J. Am. Med. Assoc.* 186: 831–834, 1963.
44. PENFIELD, W., AND H. JASPER. *Epilepsy and the Functional Anatomy of the Human Brain.* Boston: Little, Brown, 1954.
45. TOOLE, J. F. *Cerebrovascular Disorders.* New York: McGraw-Hill, 1967.

The Hospital

46. ANDERSON, O. W. *The Uneasy Equilibrium.* New Haven, Conn.: College and University Press, 1968.
47. FELIN, R. *The Doctor Shortage: An Economic Diagnosis.* Washington, D. C.: The Brookings Institution, 1968.
48. FREIDSON, E. *The Hospital in Modern Society.* Cambridge: Harvard Univ. Press, 1969.

Ophthalmology

49. BECKER, B., AND R. N. SCHAFFER. *Diagnosis and Therapy of the Glaucomas.* St. Louis: Mosby, 1965.
50. HOGAN, M. J., AND L. E. ZIMMERMAN. *Ophthalmic Pathology, an Atlas and Textbook.* Philadelphia: Saunders, 1962.

Otolaryngology

51. BALLENGER, J. J. *Disease of the Nose, Throat and Ear.* Philadelphia: Lea & Febiger, 1969.
52. VON BEKESY, G. V. The Ear. *Sci. Am.* 197: 66–78, 1957.

Radiology

53. GRIGG, E. R. N. *The Trail of the Invisible Light.* Springfield, Ill.: Thomas, 1965.

Rehabilitation Medicine

54. LICHT, S. (editor). *Rehabilitation and Medicine.* New Haven, Conn.: Elizabeth Licht, Publ., 1968.
55. SUSSMAN, M. B. (editor). *Sociology and Rehabilitation.* Washington, D. C.: Am. Sociological Assoc., 1966.

Chapter IV

Dental science

JOSEPH F. VOLKER, D D.S., Ph.D.

Dental science is concerned with health and disease of the teeth (dentition), and gums (gingiva), the underlying bone, and the structures in and about the oral cavity that may influence the course of systemic disease. The common dental afflictions are tooth decay (dental caries) and disease of the soft tissues and bone that support the teeth (periodontal disease). These two diseases have plagued the majority of the world population. Reliable estimates indicate that, in the United States, there are nearly a billion unfilled, decayed teeth; that in 45- to 54-year-old adults, 85% of the men and 74% of the women have periodontal disease, and that a surprisingly high percentage (45%) of persons 60 and older have lost all their natural teeth (edentulous).

Achievements in the basic sciences in dentistry are intimately interwoven with medical advancements and in most instances indistinguishable from them. For this reason these achievements will not be repeated here. Only those that have application to dental practice will be considered. This chapter is concerned with: 1) prevention of dental caries (decay) and periodontal disease; 2) diagnosis, prevention, and treatment of malalignments of the teeth and jaws (malocclusion); 3) treatment, restoration, and replacement of teeth; 4) removal of teeth and related diseased parts; and 5) restoration of destroyed parts of the jaws and face (surgical prosthesis).

During the past 30 years, significant progress has been made in dental research focused on all aspects of lifetime dental health of the individual. This chapter discusses these major areas of progress, excluding subjects such as anesthesiology and antibiotics, which are contained in the other medical and health sciences.

DENTAL DECAY

Tooth decay is the most rampant disease affecting mankind. It is intensely destructive and few individuals escape its consequences (Fig. 1). In this very prevalent disease, lesions are never self-healing but must be repaired and restored by artificial means; hence, prevention is of prime importance. While scientists are searching for better preventive and restorative measures and for improved dental materials, practicing dentists continue

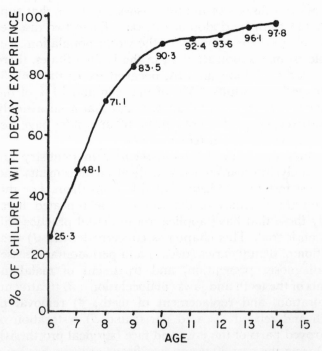

FIG. 1. Percent of children at each age from 6 through 14 years with decay experience.

to be overwhelmed with the task of repairing damage caused by decay and other dental disorders.

Three requirements are considered essential for the production of tooth decay: *1*) bacteria, *2*) the presence of certain food in contact with the teeth, specifically foods containing sugar or other carbohydrates, and *3*) a susceptible tooth, i.e., one that cannot withstand the onslaught of the decay process.

One of the significant achievements of the last few decades is the proof that microorganisms (bacteria) are necessary for the initiation and spread of tooth decay. Such proof was made possible by the development of a method of raising and maintaining germfree animals. The teeth of these animals do not decay so long as they remain free of specific bacteria. But when these animals cease to be germfree, either naturally or by intentional introduction of bacteria, their teeth do begin to decay. Certain bacteria are more prone to cause tooth decay than others. Germs from animals with tooth decay are effective in causing tooth decay in germfree animals, as are certain kinds of streptococci from human beings. These findings supplied initial evidence that tooth decay is a transmissible disease and that microorganisms (bacteria) are indispensable to its development. This highly significant discovery raises the hope that in the not too distant future, prevention of dental decay may be achieved by immunological methods, that is, by increasing the body's ability to destroy the bacteria known to produce tooth decay.

To begin the destructive process, these bacteria must have food on which to grow, the kind of food that is effective in helping to bring about the decay of teeth. A diet capable of causing tooth decay will not do so if the food is put directly into the stomach through a tube that bypasses the mouth. Furthermore, it has been found that table sugar (sucrose) is more disposed to initiate decay than other sugars and starches, probably because its presence is conducive to the formation of heavy plaque on the teeth. Accumulating on the surfaces of the teeth, plaque is a gelatinous mass of microorganisms bound together by certain insoluble and modified forms of sugar that are built into long chemical chains by the action of bacterial enzymes on sucrose (table sugar). The bacteria whose enzymes bring about the chemical change that results in plaque formation have repeatedly produced tooth decay in experimental animals. Plaque clings tenaciously to the enamel surfaces of the teeth and cannot be removed by merely rinsing the mouth; toothbrushing and use of dental floss are necessary. However, neither toothbrushing nor dental floss adequately cleans plaque from such imperfections in tooth surfaces as the pits and fissures. These surface areas are where decay starts early. The bacteria live and multiply within plaque material, converting the sugars into organic acids, which in turn cause decay in the enamel surface at the interface between the plaque and the tooth enamel. This interfacial area is of grave importance to those studying this disease. Plaque formation constitutes a problem on which researchers are exerting much time and effort in seeking ways to prevent and destroy plaque with enzymes, antibiotics, or antiseptics. Clinical studies with this as the objective are being undertaken.

For some time, commonly prescribed antibiotics have been used to destroy microorganisms that cause tooth decay. A drawback to their use is the possibility of developing bacteria that become resistant to antibiotics commonly administered to combat acute and dangerous illnesses, which would therefore be ineffectual if they were urgently needed for acute infections. There is also fear of developing an excessive growth of disease-producing fungi, because the antibiotic could destroy the natural balance of the oral microbial environment. There are many antibiotics that the medical profession is not presently using. These antibiotics are being reevaluated as possible preventives of human dental decay. Actinobolin appears promising, because it shows striking reductions of decay in experimental animals but does not upset the oral flora balance and it is not absorbed to any great extent from the gut.

For tooth decay, there must be a susceptible tooth as well as bacteria and sugary foods. A person's natural resistance to tooth decay is therefore important. The

longer a tooth remains in the mouth, the more it ma-
tures and the less vulnerable it becomes. This is because
certain chemical changes that occur on the enamel surface make it harder (more mineralized) and better able
to withstand the decay caused by the acids formed from
the action of bacteria on sugar. At the interface between
plaque and enamel, a kind of constant battle is waged
by forces that make teeth softer and more susceptible
to decay (demineralization) versus the forces that make
teeth harder and more resistant (mineralization). Scientists have sought ways of hastening tooth maturation
to speedup development of natural resistance to decay,
and to find a means of helping the tooth to combat demineralization (decalcification), which makes teeth less
resistant. In this research a significant scientific breakthrough occurred in connection with the use of fluorides.

The prevention of dental caries by the use of fluorides
is one of the outstanding achievements of our time. Not
only did the development of the fluoride–caries theory
provide the energy and means for expanded research
efforts in prevention, but community water fluoridation,
which required acceptance by the public, introduced
the dental research investigator into the fields of the
social sciences and public relations.

It was observed that in communities where water
contained approximately 1 part per million (1 ppm)
of fluoride, the persons drinking this water from birth
had 60% less dental decay than persons who drank
nonfluoridated water for an equivalent period. Furthermore, researchers also observed that no harmful effects,
such as dental fluorosis, resulted where there was 1 ppm
of fluoride in the water. Early confirmation of these
findings suggested that artificial water fluoridation could
be used to reduce dental caries in humans.

In the United States, both the Newburgh-Kingston,
New York, and the Grand Rapids, Michigan, studies
began in 1945 with 1 ppm of fluoride being added to
the public water systems (Fig. 2). In the New York State
study, a sample population of children, from both the
test and control communities, received complete physical
examinations that included X-raying the leg bones. No

harmful effects from water fluoridation were noted in this investigation, nor have any adverse effects from artificial fluoridation ever been reliably documented. In the past 25 years, findings from subsequent studies on water fluoridation have all verified the results of the early studies.

These experiments were more than research efforts, they served as important demonstrations to the public and to the scientific world that a simple and entirely

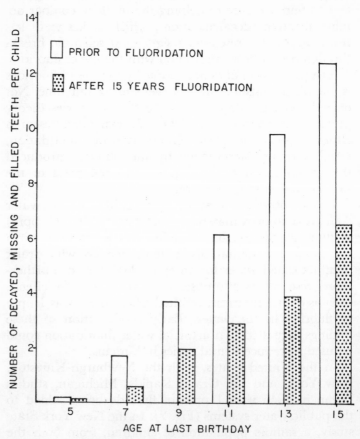

FIG. 2. Comparison of dental caries prevalence in Grand Rapids, Michigan, following 15-year fluoridation period. (Modified from *Fluoridation* by Frank J. McClure.)

practical method had been developed for reducing dental decay within the total population. Most important, the method was devoid of harmful effects.

Slowly but steadily communities have adopted water fluoridation. There has been, as with many public health measures, a belligerent and persistent opposition, with attempts to prevent progress by scare tactics and legal means. Court decisions have upheld the legality of this public health measure. Similar resistance was evident previously with other public health measures: milk pasteurization, vaccination for smallpox, and water chlorination. Despite organized opposition, over 80 million of the United States population of 204 million are now drinking artificially fluoridated water, while another 8 million drink natural fluoride water. At present, six states have mandatory water fluoridation laws. It is imperative to the welfare of dental health that effort be continued for community water fluoridation.

Water fluoridation is superior to other methods of fluoride treatment, since no trained personnel are required to give the treatment, nor is any extra effort required by those benefited. Considering water fluoridation from an economic point of view, the findings are very impressive. The time required of the dentist for corrective care and the cost per child is half of that necessary in nonfluoride areas. For the 88 million with communal fluoride water supplies the reduction in suffering and the financial savings are too vast to estimate accurately. The reduction in individual need for dental care in the fluoridated area allows dental treatment for a larger number of the population in the same amount of time previously required for a smaller number. Premature loss of primary (baby) teeth resulting from dental decay is one of the major causes of crooked teeth in children. Children drinking fluoridated water have only half as many tooth extractions as those in nonfluoridated communities. Consequently there are less malocclusions and less need for preventive and corrective orthodontics.

In areas where communal water supplies do not exist, school water supplies can be treated so that the fluoride intake per child, from the school drinking fountains, will satisfy the daily required amounts to prevent tooth decay. Fluoride tablets have been shown to be effective in nonfluoride areas if given during the years of tooth development, from birth to 14 years of age.

The fluoridation of table salt and of milk is now under experimentation. Adding fluorides to table salt has proved successful in other countries where public water supplies are limited.

Early work with fluorides indicated that either naturally fluorosed teeth or teeth treated with fluorides had a reduced solubility in acids. Since the acid theory of dental caries had been widely accepted by most investigators, it became a real challenge to find a method of applying fluorides on tooth enamel to aid in resisting decalcification and dental caries. A search for other means of reducing dental decay became important, because while water fluoridation could reduce dental caries by 60% the remaining 40% needed to be prevented by other means. For the large segment of the United States population who do not consume communal water, some other means of fluoride supplementation had to be developed.

Application of different fluoride concentrations to the tooth surface has been achieved by several methods.

Prophylactic pastes. Although prophylactic pastes containing fluoride will reduce dental decay significantly, it is questionable whether they are an equally satisfactory substitute for topical fluoride applications.

Fluoride dentifrices. Fluoride toothpastes containing either stannous fluoride, acidulated phosphate-fluoride, or sodium monofluorophosphate, when used as directed, have shown a greater reduction in dental caries than have dentifrices without the added fluoride. Positive reductions have been shown to result from use of fluoride toothpaste, but because of the wide variation in the amount of caries reduction reported from the different clinical studies, a cautious interpretation must be given as to the superiority of any specific dentifrice.

Fluoride mouthwashes. In studies conducted in the United States and abroad, fluoride mouthwashes appear to be effective. Studies are now in progress in the United States on concentrations of fluorides safe for home use.

One promising development, which recently has been tested, is to prevent tooth decay by adding chemicals to sugary foods and confections during their manufacture, to reduce or eliminate their capacity to invoke tooth decay. Animal studies have shown phosphates, in certain chemical combinations, are capable of reducing tooth decay by local action within the mouth. Some phosphates have proved more effective than others. Future clinical trials from this approach should benefit humans. One phosphate incorporated with sugar in chewing gum produces a significant reduction in tooth decay.

There is a need for more clinical trials on the usefulness of certain phosphates added to sugar-coated cereals. Complete studies have yielded divergent results. One thing is clear, the more frequently sweets are eaten and the longer they remain in the mouth, the more tooth decay results. In fact, it is entirely possible that decay could be prevented by eliminating all carbohydrates from the diet. This solution is not practical for long periods of time, because people are not likely to refrain from carbohydrate consumption. However, in-between meal eating of sugar-containing foods could be easily avoided.

Another recent and important development in preventing dental decay is the use of plastic materials to seal the pits and fissures on tooth surfaces, especially the surfaces that make contact in biting and chewing. Now, with the availability of new and improved plastic materials, these surface defects can be sealed without drilling and without destroying the surface of the tooth.

PERIODONTAL DISEASE

Periodontal disease, another widely distributed scourge of man, is a progressive disease of the periodontal tissues (gums) and the bone that support the teeth. Insidious in its onset, steadily moving along its destructive course, the unwary victim is usually alerted by the loosening of one or more of his teeth, which he may be in danger of losing. Frequently the other teeth may be doomed because the destructive process has progressed beyond the point of effective treatment. This disease is the greatest single cause of tooth loss, particularly of that most distressing disaster, the loss of all the teeth.

Despite tremendous effort and the use of every weapon of the dental scientist, the way in which this elusive disease originates, and its essential pathogenesis, are not yet completely understood. Therefore, the dentist cannot attack the disease on a public health basis; he can only strive to eliminate or minimize the damage. He must attempt to motivate the patient to employ methods and techniques of dental care that many patients may be too careless or disinterested to perform adequately. Essential progress in control of this disease has been made through patient–dentist collaboration on preventive measures. These will not completely control this disease in general population, but they are the only ones available.

Fortunately, most people in our society are interested in and receptive to the teaching and enforcement of the minimal regimen necessary for good results. These measures begin with the dentist's removal (debridement) from the teeth of harmful material (such as plaque and calculus). Subsequently, the patient's efforts to personally continue this type of procedure at home will assist in controlling the problem. In the early days (nearly a century ago), debridement consisted largely of removing calculus from the teeth. Unfortunately, in the treatment of the general population, there has been little progress beyond that point. For the best results from use of available measures of control close cooperation between dentist and patient is necessary. Because of difficulty in securing complete cooperation from the patient, this is one of the weakest points in the management of this disease.

In addition to the basic debridement technique, there is a wide range of surgical procedures, some of which eliminate the abnormal spaces or pockets that develop between the teeth and supporting bone and soft tissues (gums) so that the patient can effectively manage his oral hygiene without constant professional assistance. After removal of plaque and calculus (debridement), the goal has long been to encourage the natural repair of the tissue damaged by the disease.

About 20 years ago a major advance was realized, when it was proved possible to reverse the destructive process of the disease and foster actual regrowth, nat- ural repair, and rebuilding of lost bone, dental cement, and periodontal tissue. This accomplishment was possible in a small proportion of instances in which the conditions at the damaged site particularly favored natural tissue repair. Currently dental scientists seek to discover ways of inducing tissue repair at sites not so favorably disposed to natural repair.

Inserts and grafts of various kinds, including pieces of bone, bone dust mixed with blood, plaster, cartilage, and fresh and frozen bone marrow, have been used in an effort to induce the growth of new bone and tissue to restore the supportive attachment to the tooth that has been denuded by disease. So far, the results have not been clear-cut but the successes have been numerous enough to spur investigators to further efforts.

Pockets adjacent to the teeth are not the sole problem confronting the periodontists. The quality and arrangement of gum tissue are highly important, particularly from the standpoint of disease prevention. In this instance, grafts of the patient's own oral tissue, as from the palate, have been so successful in correcting these faulty tissue arrangements that this procedure has become routine dental practice.

In many patients periodontal disease is so advanced that special restorative methods are necessary. To meet these, a small, highly skilled group of restorative workers established the subspecialty of Periodontal Prosthesis. The relegation of a patient to full dentures carries a connotation of professional failure. These periodontal prosthetists, however, are limited in number and their arduous task is very time-consuming. They can treat very few patients each year. Consequently, the dental scientist is thrown back to the search for simpler methods, preferably of prevention, and an effective "public health" approach.

This search continues on two fronts: *1)* adequate prevention and control of the disease itself, and *2)* more effective methods of clinical management, especially techniques of repair and induction of natural repair processes. Calculus formation was originally considered the chief agent associated with the onset and development of periodontal disease. Researchers now feel that plaque formed on the teeth (discussed in connection with tooth decay) is the prime irritational factor. Thus, efforts to prevent the disease and maintain healthy teeth have shifted from calculus to plaque control, and methods of patient tooth care have come to stress more scrubbing and less massage. In the past 25 years epidemiological surveys have shown correlation between this disease and plaque formation, and experiments have shown a relationship between plaque accumulation and disease of the gums.

Many studies have shown that good home care plus rather basic periodontal treatment procedures can substantially reduce the disease and consequent loss of teeth. Methods of staining plaque on teeth have come into use to enable patients to determine how well they clean their teeth, thereby improving home care. Action by some bacteria has been implicated in plaque formation. It is believed that enzymes associated with these bacteria exert a powerful influence on the progress of destructive periodontal disease.

OTHER CLINICAL CONSIDERATIONS

Diseases of the teeth and other oral tissues may produce excruciating pain; however, pain of dental origin often goes beyond a carious or periodontally infected dentition. Pain, instead, may be suggestive of pathology of greater significance and on occasion relate to a sinister or terminal disease.

Malocclusion, the improper meeting of the upper and lower teeth, or the upper and lower jaws, is frequently the source of discomfort for the patient who complains of pain in the temporomandibular joint or preauricular area. However, studies have provided data which demonstrate that numerous other ailments may be associated with this disorder, such as diseases of the central nervous system, of the ear, nose and throat, the arthritides (particularly rheumatoid arthritis), tumors and cancers of the associated glands and soft tissues, and occasionally disorders stemming from a psychogenic problem.

Advances in radiology, particularly those X-ray studies recorded from a predetermined tissue level (tomography) and X-ray motion pictures (cineradiography) utilizing opaque material and fluoroscopy, have provided important information in diagnosis and treatment of temporomandibular joint disorders.

Another large group of patients suffer what is termed "atypical face pain," a distressing condition that often frustrates patient, dentist, and physician alike. Although a solution often eludes the practitioner, information gained from clinical research of the past 30 years indicates that cautious, conservative treatment, and continued observation are the best means of care.

Recently, clinical dentistry has contributed to a better understanding of the nature of another systemic disorder, "Wilson's disease," a rare inherited abnormality associated with an interference of copper metabolism. The clinical phenomena observed by dentists have been a demineralization of the jaws and a change in occlusion.

By recognizing these more serious diseases and disorders that researchers have identified, the dentist is better able to assist his patients in total health care.

DELIVERY OF DENTAL SERVICES

The sections of this report dealing with dental caries and periodontal disease make it apparent that almost every United States citizen needs dental care at some time, and that most citizens experience dental disease so severe that periodic care is needed to prevent irreparable damage to the dentition. Numerous studies relate dental needs to available dental manpower. These studies have shown that *needs* greatly exceed productivity, although *demand* for dental care does not exceed productivity by nearly so wide a margin. Awareness of this situation has led to efforts to increase the ratio of dentists to population, and to improve the productivity of the dentists themselves. The profession has worked hard to train additional dentists without success in keeping up with population growth. The record is much better in the field of productivity where large increases have been realized within the past 25 years.

In 1930, the United States population of approximately 122 million was served by 70,000 dentists, one to 1,750 people. Today there are a little more than 100,000 dentists or a ratio of one to 2,000. The ratio is inadequate since one dentist can render modern comprehensive dental care to only 1,000 patients per year.

It has taken concentrated effort to maintain even this 1-to-2,000 ratio. At the close of World War II, there were 40 dental schools; now there are 53, with several more being organized. From less than 200 dental graduates in 1930, the record has improved to 3,500 graduates in 1970. If the profession is to keep pace with the growth of the population, it must reach a figure of almost 4,500 graduates annually by 1980.

There are many new techniques that improve productivity, though most have not been adopted as widely as might be desired. The greatest improvement in dental instruments has been the development of rotary cutting tools. Air-driven turbines or ultraspeed, belt-driven mechanisms now develop 400,000 revolutions per minute (rpm), as against the 3,000 or 4,000 rpm of the conventional electric motors in use a generation ago. These new instruments, being properly cooled, reduce the preparation time for large cavities by as much as 75% and for even the simplest cavities by perhaps 50%; and perhaps the greatest benefit of all, patient discomfort is reduced. Local anesthetics have paved the way for "Quadrant Dentistry." In a very short time, at the same sitting, a group of cavities can be prepared and filled without pain, in a quarter of the mouth.

Of greater importance to the dentist than his instruments, are the trained assistants now available to him. A dentist can double his output with the help of one full-time and one half-time assistant. The full-time assistant remains beside instruments, developing X rays, seating patients, and so on. Since instruments can be handed to him, the dentist can remain seated, and thereby operate under less mental and physical strain.

Further developments of this technique have been given the term "Four-handed Dentistry," since the hands of the dentist and of the full-time chair assistant are all utilized to the greatest mechanical advantage. Old-fashioned methods of dental operating have been analyzed and found to involve instances of abnormal posture that contributed to inefficiency and fatigue on the part of the dentist. Dental equipment has been redesigned, with the patient in a supine postion and the dentist seated so that his back is straight and his head relatively erect. The assistant is seated opposite the dentist so that she can reach the field of operation easily and manipulate various hose and wire attachments, the instrument tray, and the assorted materials without leaning, twisting, or overextending her arms.

Dental units were redesigned to bring essential pieces of equipment within the direct reach of the dental team. High-velocity suction was developed to eliminate the time-taking use of the cuspidor and to permit use of the coolants for high-speed cutting tools. Trays were designed for holding all the instruments required for specific types of dental operating; this saved numerous entries into the dental instrument cabinet during each operation. These developments have increased the output of the individual dentist above the levels reached from acquisition of more auxiliary personnel; and probably have improved his endurance, the accuracy of his work, and the span of his working life.

The final thrust in the use of auxiliary personnel has been the delegation to specifically trained auxiliaries, under the direct supervision of the dentist, of expanded duties that do not require expert professional judgment and that can be "reversed" or done over without damage to the patient. These duties include placing and removing temporary restorations, placing and carving of amalgam restorations in previously prepared teeth, placing acrylic restorations in previously prepared teeth, and finishing and polishing various types of restorations. For several years, experiments have been underway in the delegation of these duties to auxiliaries.

In programs conducted by scientists in the Department

of the Navy and the Public Health Service, dental assistants help the dentist prepare the cavities and therefore know exactly what problems he has encountered. The patient, seeing that the assistant has this knowledge, realizes that his dental repair is being performed by a team with the dentist in command. Patient acceptance of this operational procedure and the quality of the dental restorations have been excellent.

The efficiency that results from delegating expanded duties to auxiliary personnel has been adequately demonstrated under experimental conditions; the next steps are legal authorization of such delegation and its adoption by the dental profession. The American Dental Association, originally conservative toward this delegation of duties, has liberalized its attitude. The Association is now urging changes in dental practice acts and in state dental examining board regulations that will permit participation of auxiliary personnel. A few states, notably Pennsylvania and Minnesota, have adopted regulations sanctioning extended use of such auxiliary personnel. In the city of Philadelphia, specially trained assistants are operating on a four–to–one dentist ratio in a clinic maintained by the city health department. Reports from Philadelphia indicate this service has been well accepted.

The delivery of dental care to the public requires consideration of the consumer and his environment and desires as well as of the dental operating team. Large low income population groups are often neither financially able nor culturally adapted to avail themselves of private office dental practice. Dental care must be brought to the ghetto populations in their own home areas, and they respond best when they participate in the planning and management of the health services they are to receive. Among the poor there is an inability to make use of distant resources, even if financial barriers are removed. Low income people, moreover, seem to relate better to an institutional team than to a single practitioner of a cultural level different from their own. The modern trend is to treat whole families rather than certain age or sex groups.

The neighborhood health center has evolved from the charity clinics of the early 1900's. These early clinics served special population groups, chiefly children, and were paternalistic in their approach to "poor" people. The new neighborhood health centers serve entire families and look beyond strictly health matters toward general social rehabilitation through personal health services for a geographically restricted population. In the new neighborhood clinic and the area served, health center workers help to adapt health services to the needs of the consumer, and give him a sense of participation. Local people can be employed in various positions at such health centers, with economic as well as health benefit to the community.

DENTAL MATERIALS

The American people cannot expect good dental health "across the board" unless two developments are realized: 1) further major improvements in the delivery of dental care, and 2) a drastic reduction in the load of dental disease through preventive measures.

Along with the increased productivity of the dental team has been the development and improvement of dental materials. Traditional restorative materials often lacked permanency; the repair of dental restorations required as much of the dentist's time as putting in new fillings, bridges, or dentures. One of the important goals in dental science is to improve the durability of materials to be used in fillings and other restorations. These materials must be durable enough to withstand the tremendously high forces (30,000 psi, or more) exerted on the teeth and the dental restorations during biting and chewing, the great temperature changes (many exceed 100 F) of the teeth caused by foods, such as hot coffee and ice cream, and the wide differences in acidity and alkalinity caused by different beverages and foods. As important and perhaps even more critical is what is called the "microleakage phenomena." This is the slow continuous seepage of saliva, acids, bacteria, and other "debris" at the areas of contact between the enamel and the dental cement or filling material. Such leakage is one of the major causes of the recurrence of decay. Since the tooth is wet, not homogeneous in composition and microscopically dirty, it is difficult to produce a true adhesive bond between the enamel–dentin surfaces and a dental cement or filling material that is sufficient to prevent leakage. The oral cavity is an ideal environment for the repeated occurrence of corrosion, decay, and discoloration.

In the struggle to eliminate such failure, an essentially new discipline has been created—the science of dental materials. Over the past 25 years, one-third of dental research has been focused on this field. No other single factor has been more responsible for the high standard of dentistry in this country than the improvement in the physical and chemical properties of dental materials.

Research in progress may provide greater permanency of restored teeth. The basic guide for these improvements has been the program of specifications which was begun in 1928 by the joint research efforts of the American Dental Association and the National Bureau of Standards. Most of the new and improved dental materials in use today adhere to these detailed specifications.

Among the achievements of this dental materials program is the exceptional improvement of dental adhesives, which now make it possible to use resin in many instances. Resins can be more easily manipulated, and those now available resist fracture as well as dissolution by oral fluids. New composite dental resins may soon replace metallic restorations. The impressive physical test results of some are presently being evaluated; they hold great promise of providing superior dental restorations. Some of the newer, more adhesive resins no doubt will extend the lifetime of the average dental restoration even further and will minimize the leakage. Within 5 years such adhesive filling materials will be used routinely by the dentist. They may very well be used in orthodontic therapy in connection with improved orthodontic appliances that can be cemented directly to the enamel surface of the teeth, in lieu of the bands now placed on each tooth to hold the appliance. If the new adhesive

cements prove useful, the orthodontic appliances would
be more pleasing, more quickly put in place, and possibly reduce their cost. It may be that such adhesive,
film-forming materials will be perfected so that it will
be feasible to paint them on the entire outer surface
of the teeth, thus providing a barrier against the formation of plaque and calculus, and acting as a preventive
against tooth decay and periodontal disease.

Jaw–face (maxillo-facial) appliances are now being
used that would have been impossible to make 20 years
ago. The congenital cleft palate can now be repaired
with materials that provide the accuracy and stability
necessary for proper functioning. No longer do the older
unsightly resins have to be used in making artificial
organs, such as palate or nose, for the patient who has
lost the natural organs as a result of cancer. The new
materials can be fabricated into such devices, particularly those involving tissue implants that fit precisely
and are extremely well tolerated by the patient.

Research on bio-dental materials makes continuing
contribution to advancements in dental science. The
commonly used temporary filling materials lack necessary biological characteristics and are not sufficiently
tough for long duration. New materials with more suitable properties have been developed, laboratory tested,
and clinically evaluated; the results indicate that these
materials may last for at least 3 years, and they offer
the advantages of time saved and added comfort in
dealing with large population groups. It is possible they
could help solve a difficult Armed Services dental management problem that arises from a low dentist–patient
ratio, and the necessity of providing dental care for the
great number of recruits that have a high incidence of
tooth decay and periodontal disease. With the new filling
materials, it should be possible to provide stronger and
longer lasting temporary fillings that would delay further
tooth destruction until permanent restorations could
be provided at more convenient times. Some of the less
complicated manipulation and placement could be done
by unskilled personnel, under certain field conditions.
Such temporary holding material.could be useful also

in civilian dental care of children with rampant tooth decay, and could be part of a dental kit carried by astronauts on an extended trip in space.

A comparison between materials in use 25 years ago and the materials available today will indicate the breakthrough that is now in progress. The old unsightly rubber (vulcanite) denture base has been replaced by natural-appearing resins. Although silver amalgam is still widely used, this material today is more resistant to corrosion, can be inserted in a cavity more easily, and is 25% stronger. Elastic impression materials that have replaced the nelastic are more accurate and simplify the technical procedure in fabricating bridges or dentures. Gold alloys used in making partial dentures have been mostly superseded by less expensive, lighter and stronger alloys. In orthodontic appliances gold wires have been replaced by stainless steel alloys which allow in some instances for more rapid movement, with fewer adjustments of teeth that are being corrected. Biological type cements, for placement over irritated pulps or for support of permanent restorations, promote repair and healing, and enable the dentist to save teeth that would have been doomed to extraction 20 years ago.

Engineers have studied oral stress patterns during chewing, leading to improvements in the preparation of cavities to be filled, to a better design of dental appliances, to prevention of appliance failure and to better protection of the supporting tissues. Ceramics, less likely to crack or fracture, are now available for denture teeth or for jacket crowns and porcelain can be fused with gold to provide stronger and more esthetic bridge designs. Dental science and clinical dentistry have been tremendously advanced in the past 30 years by such important developments in materials.

ORTHODONTICS

The orthodontist is concerned with the growth and development of the teeth and related anatomical parts as well as their arrangement in the mouth. For any movement of tooth position that is undertaken, extensive use is generally made of the cephalostat, a

type of X-ray equipment that is especially designed for
the study of the bony structures of the face and jaws.
Cephalometric tracings made from full face and side
face radiographs (X-ray pictures) of the skull provide
a means of following the growth and development of
the teeth, and of the supporting bony and soft tissues
(gums). These films also make it possible to follow the
progress of orthodontic treatment. Widespread clinical
use of cephalometrics increased after World War II.

Patients who, because of malformation or injury,
must be in either a plaster cast or a brace that has a chin
rest, may develop a severe malocclusion. To prevent
this, orthodontics promoted development of a new type
of back brace that has a throat piece instead of a chin
rest.

There are millions of less severe malocclusions that
can be either prevented or arrested. Premature loss of
primary teeth because of tooth decay is one of the major
causes of the drifting of teeth, a common contributing
factor in malocclusion. Programs for preventive treatment of malocclusion have been established for students
within many dental schools.

Defects of malocclusion also concern speech pathologists and speech therapists who, along with the dental
scientists, are seeking causes as well as methods of correction of these deformities. Orthodontists have established
simple methods for assessing malocclusions in large numbers of persons and for providing information concerning treatment.

SURGICAL ORTHODONTICS

Both orthodontists and oral surgeons are concerned
with the commonly occurring type of malocclusion in
which the lower jaw juts forward to a seriously abnormal
degree causing impairment in speech and chewing, loss
of teeth due to difficult oral hygiene, and even psychological and personality problems. During the past 30
years, major advances have been made in surgical orthodontics, in the development of various methods for the
correction of serious dental and facial deformities. When

segments of the bone supporting the teeth, and even the dental arch itself, are abnormally placed, it has become a common, well-established practice to shift them to a more normal position with predictable and satisfying results. Major surgical procedures in orthodontics have been made relatively safe by the advances in anesthesia and antibiotics, and also by the development of sophisticated cutting instruments.

A particular abnormality is that in which an excessive vertical distance separates the upper and lower jaws, a condition known as "open bite deformity." Sometimes this can be corrected by orthodontic treatment alone, however, advice of a speech therapist is often required. Recently surgical procedures have been developed for raising, lowering, advancing, or retracting sections of either the upper or the lower jaw, or of moving them sideways, to correct these open bite deformities. These and other advances in diagnosing and correcting the various types of malocclusion involve both the orthodontist and the oral surgeon, and include such things as moving an impacted tooth into an artificially created socket in the bone in order to save the tooth.

The oral surgeon, in cooperation with the prosthodontist, has developed techniques that greatly benefit persons in need of full dentures. Metal implants inserted into the jawbones support especially made dentures and offer the wearer better retention and greater ease in chewing and speaking. Another technique involves the surgical deepening of the space (sulcus) between the jawbone and the cheek to provide more bone stability for artificial dentures.

MAXILLOFACIAL PROSTHESIS

The dental scientist is also concerned with prosthetics associated with, but outside the oral cavity, including jaw–face (maxillofacial) prosthetics made necessary by injury, disease, or congenital malformations. During the past 25 years, the numerous advances made in this area were possible largely because of new and improved materials, such as those mentioned earlier, and because of a change in the management of the patient that includes rehabilitation follow-up as well as freeing him of disease. This change evolved as a result of experience with the many patients injured during World Wars I and II, and has been fostered by the availability of antibiotics and improved anesthetics, which contribute to more complete and effective surgery and the consequent need for rehabilitation. One important development in materials is the silicone implant, which has less reaction with the tissues and fluids of the body than do the other available materials. Silicones can be used to coat materials which otherwise are not acceptable to the body. It is a most successful material for implantation in soft tissues where contour and activity are both necessary, and can be used for various prosthetic purposes as follows: in the form of sponges, which must remain soft in the body; meshworks; sheets; strips; or any other geometric shape; and they can even be injected into the tissues where they can become hardened without toxicity to the tissue.

Another important achievement is the development of a technique in which a bit of the patient's own pelvic bone marrow is placed across or about an injured bone site that is in need of healing. The bone marrow on the microfilter speeds up the healing process, and the metallic implant, which gives the desired shape, contour, strength, and rigidity during healing, can be removed later on.

The supply of new dental materials is increasing rapidly and further improvements in prosthetic techniques can be expected in the future. All the new materials have led to improvements in extraoral prosthesis, but silicones, particularly, are beginning to approach the ideal desired. They can be tinted while in liquid form and molded to a predetermined shape, and very light weight silicone foam can be used in the interior of prosthetic devices. Improvements in other prosthetic materials (metallic, solid plastic, porous plastic, ceramics) are also progressing rapidly.

To meet the increasing demands for dental services,

184

CHAPTER IV

it is necessary not only to improve materials, instruments, and techniques, but also to train increasing numbers of dental scientists, specialists, and general practitioners and assistants in the most recent knowledge and techniques. Within the past 30 years, many specialties and specialists have come into the dental picture, and the art and science of dentistry has become increasingly more complex with the expansion of dental knowledge. The dental research investigator must continue in this quest for knowledge until preventive means are found for completely eliminating the dental ailments that now affect mankind.

SELECTED ADDITIONAL READING

1. ALLEN, N. D. Handicapping malocclusion assessment record in direct mouth examination. Am. J. Orthodontics 58: 67–72, 1970.
2. AMERICAN DENTAL ASSOCIATION. Guide to Dental Materials and Devices (4th ed.). 1968, Chicago, Ill.
3. BLIX, G. Nutrition and caries-prevention, symposia of the Swedish nutrition foundation III. Uppsala, Sweden: Almqvist & Wiksells, 1965.
4. BODINE, R. L., AND C. I. MOHAMMED. Histologic studies of a human mandible supporting an implant denture. J. Prosthetic Dentistry 21: 203, 1969.
5. BOUCHER, C. O. (editor). Swenson's Complete Dentures (6th ed.). St. Louis: Mosby, 1970, p. 569.
6. DEWEL, B. F. Orthodontics achievements and responsibilities. Am. J. Orthodontics 54: 823–830, 1968.
7. Evaluation of agents in the prevention of oral diseases. Ann. N. Y. Acad. Sci. 153: Art. 1, 1968.
8. FINN, S. B. Clinical Pedodontics (3rd ed.). Philadelphia and London: Saunders, 1967.
9. GIBILISCO, J. A., N. P. GOLDSTEIN AND J. G. RUSHTON. The differential diagnosis of atypical facial pain. J. Lancet 85: no. 10, 1965.
10. GOLDMAN, H. M., AND D. W. COHEN. The infrabony pocket: classification and treatment. J. Periodonol. 29: 272, 1958.
11. HARRIS, R. S. Art and Science of Dental Caries Research. New York and London: Academic, 1968.
12. HITCHCOCK, H. P. Treatment of malocclusion associated with scoliosis. Angle Orthodontist 39: 64–68, 1969.
13. LEE, T. C. An historical review of implantology. In: Oral Implantology, edited by A. N. Cranin. Springfield, Ill.: Thomas, 1970.

14. McClure, F. J. *Water Fluoridation; The Search and the Victory.* Bethesda: National Institute of Dental Research, 1970.
15. National Institute of Dental Research. *Adhesive Restorative Dental Materials.* Washington, D. C.: U. S. Govt. Printing Office, 1966, vol. ii.
16. National Institute of Dental Research. *Barnacle Cement as a Dental Restorative Adhesive,* Natl. Inst. Health Publ. no. 151, Bethesda, Md., 1968.
17. Ogle, R. G., J. A. Gibilisco, N. P. Goldstein and R. V. Randall. Oral roentgenographic changes in Wilson's disease. *J. Lancet* 87: no. 12, 1967.
18. Phillips, R. W., and G. Ryge (editors). *Adhesive Restorative Materials.* Spencer, Ind.: Owen Litho Service, 1961.
19. Prichard, J. F. The intrabony technique as a predictable procedure. *J. Periodontol.* 28: 202, 1957.
20. Richardson, S. A. Some social psychological consequences of handicapping. *Pediatrics* 32: 291–297, 1963.
21. Smith, D. P., L. F. Pilling, J. S. Pearson, J. A. Gibilisco, J. G. Rushton and N. P. Goldstein. A psychiatric study of atypical facial pain. *Can. Med. Assoc. J.* 100: 286–291, 1969.
22. Sweeney, W. T., G. M. Brauer and I. C. Schoonover. Crazing of acrylic resins. *J. Dent. Res.* 34: 306–312, 1955.
23. Sullivan, H. C., and J. H. Atkins. Free autogenous gingival grafts. I. Principles of successful grafting. *Periodontics* 6: 121, 1968.

Chapter V

Food

EMIL M. MRAK, Ph.D.

Food is a most basic and important component of man's life. The biomedical and allied sciences have been focused on preventing and controlling disease, extending life and alleviating human misery. This emphasis on enhancing man's health and well-being has resulted in increasing numbers of people and people who live longer with a consequent need for increased production of nutritious food. Although recently there has been increased scientific research and social emphasis on population control, the most optimistic estimates project an increase in world population of at least 40 % by 1985 and some demographers concede the possibility of an increase of over 50 % (3.3 billion to 5.0 billion). By the year 2000 the world's population likely will range from 6.0 billion to 7.15 billion. By then man must find more effective ways of feeding himself, regulating his population, and stabilizing his physical, social, and economic environments. This can be achieved only through accelerated scientific research.

Many disciplines within the biological and physical sciences have been applied systematically by man in agricultural research to develop the technology required to meet constantly increasing food and fiber needs. United States scientists in universities, the U.S. Department of Agriculture (USDA), and industry have blazed new trails in these supporting fields including genetics, plant breeding, plant physiology, plant nutrition, and soil–plant–water relationships. Engineers and chemists have likewise had very important supporting roles. Agriculture, however, will not be able to indefinitely meet ever-increasing food and fiber demands since elimination of hunger depends more on reducing the population increase (a geometrical process) than on increasing food production (an arithmetical process).

The arable land in the world is limited. As of 1965, there were an estimated 2.4 potential arable acres per person in the world, as contrasted with 1.0 acre then producing food, feed, and fiber. For North America, the estimate was 4.5 and 2.3 acres, respectively.

In 1964, the 48 states had 444 million acres total in cropland, of which 292 million acres (66 %) were harvested for crop production. Agricultural efficiency has increased and total cropland acreage has decreased from the 478 million acres used in 1950. Of total U.S. crop acreage in 1964, 37.5 million acres were irrigated; in 1970 the 59 crops regularly reported on by the USDA were harvested from 289,496,000 acres.

Through the application of scientific research and technology today's American farmer produces enough food to feed himself and 42 other people including 5 people living outside the United States. In contrast, the average U.S. farmer a century ago supplied only enough farm products to feed himself and fewer than 5 other people. Because our agriculture has become so efficient, people in the U.S. have time to do many things besides produce food for themselves or to earn the money to purchase such food. Without the agricultural advances triggered by research that have freed people from the drudgery of limited production of the land, there would be little labor available to man the factories, stores, museums, and all the other endeavors that made our modern lives rewarding and satisfying. Yet the average citizen is unaware that today he works less than 1 day out of a 5-day work week to get the food that his family needs. This means that each of us has 4 days of a 5-day work week available to obtain income that provides housing, transportation, schooling, medical care, and all the other things that contribute to our comfort and leisure.

Scientists and farmers in America, through cooperative development and application of new biological
and physical knowledge, have freed this nation from the
possible horrors of famine and enabled it to take its
place among the leaders of the world. Many scientific
and technological advances have contributed to this
food production revolution that enables people of the
United States to have a bountiful food supply that is
more nutritious, more varied, and less expensive than
that of any other nation in the world at this time or any
time before. This phenomenal increase in labor efficiency is due largely to the improved agricultural technology initiated during the past 30 years. This "Green
Revolution," which began in the U.S., has been made
possible by biological research conducted prior to and
intensified during these last 35 years.

The early research and experiments, which started
to change farming from a way of life to an industry,
were important but they cannot begin to match the
advances during the last 35 years. Many of the early
day developments were in the art of agriculture rather
than the science of agriculture. They concerned themselves with how to grow crops—what to plant, when to
plant, and how to plant, grow, harvest and store.

The truly major advances have come from the application of knowledge developed in studies on how crops
grow. The development and introduction of the new
knowledge accumulated from the basic sciences, such
as mathematics, physics, chemistry, botany, zoology,
bacteriology, and genetics, have been through research
applied not only to the growing of crops and animals,
but also to handling, processing, and utilization of plants
for human food.

FOOD FROM PLANTS

The understanding and application of the science of
genetics has enabled man to develop new varieties of
plants through plant breeding that produce the food we
want, where we want it, and when we want it in the
most economical fashion. In turn, the understanding
of plant physiology and plant nutrition has enabled

man to provide the needed nutritional qualities in the needed quantities. The genetic, zoological, and chemical sciences have contributed to man's ability to control food supply reducing crop pests—insects, nematodes, rodents, diseases, and weeds—that interfere with the efficient production of food from plants.

The mechanization of agriculture, especially of the harvesting processes, has played an important role in keeping food costs low. It is an important economic fact that the food cost to Americans, in terms of its percentage of their income, is now 18%, the lowest in the world. With the development of varieties of plant types more adapted to harvest mechanization and the creation of many new harvesting devices for handling highly perishable fruits and vegetables, as well as nonperishable crops, has come a resulting reduction in the required farm labor forces to the present low of 5% of the U.S. population. Research designed to reduce the backbreaking stoop-labor requirement for vegetable production has been largely responsible for the reduction from 533 million man-hours in 1954 to 439 million man-hours in 1968, a reduction of about 17.6%.

Increasing government funds for basic research in scientific disciplines in the period 1950-1965 resulted in discoveries of several basic mechanisms involved in plant growth and development. These include discoveries in the basic biochemical processes of photosynthesis—how plants manufacture food substances from water, carbon dioxide, and other essential elements. Plant biochemists and plant physiologists have unraveled the mysteries of many of the critical steps in photosynthesis and how light is involved in the control of flowering, dormancy, formation of tubers, and seed formation. In the photoperiodic response, for example, infrared light has been found to have the same effect as a long dark period. The partial understanding of the photosynthetic mechanism, the harnessing of the sun's energy into food for man, on which all earth life depends, and of the light-sensitive pigment phytochrome and its role in the utilization of light reactions rank among the milestones of the current era. These, without doubt, foretell

of additional invaluable technological advances in the
plant sciences.

Understanding water needs of plants and development of improved irrigation practices in meeting crop needs have contributed significantly to both food quantity and quality. Increased knowledge of moisture and salt movement in soils and of crop response to water stress has greatly increased irrigation efficiency and reduced the problem of salinization of irrigated areas. Crop land under irrigation increased in the United States from approximately 18 million acres in 1940 to 37.5 million acres in 1964. The increasing use of irrigation has helped immeasurably in increasing the availability of specialty vegetable and fruit crops.

Developments in biochemical genetics have been applied to breeding superior plant varieties. The recent discovery of the "Genetic Code" ranks as a major achievement of biology since it explains the central mechanism of heredity, evolution, and life. As noted in the chapter on basic biomedicine, this research has revealed how the sequence of bases in deoxyribonucleic acid (DNA) spells out instructions which are transcribed into ribonucleic acid (RNA) for subsequent translation into proteins by means of the code. The utilization of "hybrid vigor" beginning in 1916 by scientists of the Connecticut Agricultural Experiment Station, which revolutionized corn production by 1941, has been extended to other grains, sunflowers, and to several vegetable crops, especially tomatoes, onions, melons, spinach, cucumbers, and some crucifers.

The genetic trait for male sterility is heritable; thus, the plant breeder can produce a male sterile type of plant and grow large populations of these as the female parent. Another variety of the plant is selected as the male parent. The controlled pollination of the two parent plants with desired characteristics results in superior hybrid crop plants. Prior to the use of the "male sterility" characteristic in plant breeding, pollen-producing structures in flowers of the female parent plants were removed by hand. Thus, growing large acreages of hybrid seed plants was too expensive or time consuming. In addition, the discovery of the fertility-restoring factors in the bread wheat and the trisomic mechanism in barley in the early 1960's stimulated the research efforts to develop new hybrid wheat and barley varieties.

Many farm crops, including vegetables, have been improved genetically through interspecific hybridization, with the recent discovery of methods to overcome severe compatibility barriers that often prevail between wild and cultivated species. These methods; developed for the most part since 1950, include the altering of chromosome numbers of parental species (polyploidy), use of genetically diverse intervarietal hybrids as parents, use of bridging species, and laboratory culture of plant embryos. Radiation-induced mutations have led to considerable progress in breeding in recent years.

Great strides have also been made in breeding crop plants that mature faster, are adapted to a wider range of conditions, and that grow to a uniform size and ripen at the same time so that they can be machine harvested. Higher levels of resistance to the major diseases and insect pests are continually being bred into agronomic and horticultural crops thereby improving the quality of our environment by reducing the need for pesticide use. The last 15 years have seen the improvement of three kinds of cereal crops that have greatly increased the world's food supply, and hold promise for making significant contributions in the future. These are dwarf wheat, dwarf rice, and high lysine corn.

Wheat and rice provide about 41% of man's total daily caloric intake. Also important for many in the battle for caloric adequacy are corn (5.5%), potatoes (5.0%), and sorghums plus millets (4.0%). Nearly three-fourths of the world's harvested crops are the grains wheat, rice, corn, millet, sorghums, barley, oats, and rye. Thus, agronomic research has been focused on these crops—and with singular success.

The widely acclaimed "Green Revolution" currently underway in developing countries has resulted in phenomenal increases in food grains (especially wheat and rice) in many heavily populated countries such as Mexico, India, Pakistan and the Philippines. This rapid

progress in feeding hungry people was made possible
by locally applying technology developed previously
by biological scientists from "developed" countries, particularly the United States. Many of the native scientists
in countries with whom the U.S. scientists are cooperat-
ing, received their training in the U.S. universities as
undergraduates or graduate students. The Rockefeller
and Ford Foundations have had an important and effective role in the worldwide "Green Revolution"
through their international agricultural research centers
such as International Maize and Wheat Improvement
Center in Mexico and the International Rice Research
Institute in the Philippines, which initially were primarily staffed by U.S. agricultural scientists dedicated
to international service.

By combining "dwarf" genes available in the world's
wheat and rice gene pools with tall varieties having good
commercial qualities, the Rockefeller Foundation plant
breeders have developed numerous exceptionally high
yielding dwarf wheat and rice varieties. The strong
stemmed, short stature varieties are highly responsive
to the most modern cultural technology, including
greater rates of fertilizer and improved irrigation practices because they do not grow tall and fall to the ground
as the standard varieties do. Moreover, the leaves of
many of the new varieties are erect, thus improving the
plant's ability to intercept more sunlight for photosynthesis.

Another "breakthrough" in crop improvement
through genetics and plant breeding is the production of grain crops with genetically changed qualities
needed for better human nutrition. High lysine corn
is an example. For a long time agronomists and nutritionists have realized that the highly productive corn
hybrids of modern agriculture are less nutritious than
they could be. Because of low lysine content the quality
of their protein is reduced. In countries having a low
meat or protein diet, the improved higher lysine containing varieties can be used as a means of helping to
overcome the protein deficiency disease known as kwashiorkor.

In 1963, a team of scientists at Purdue University published the results of research showing that a well-known mutant of corn, opaque-2, which was present in their genetic collection, had twice as much of the amino acid, lysine, as standard corn. This increased lysine content was easily transferred to desirable, high yielding hybrids. Geneticists with the Colombian Agricultural Program of the Rockefeller Foundation realized the potential of this genetic strain for increasing the food value of corn in the tropics. They conducted experiments with undernourished children in Colombia, which demonstrated that the effects of protein deficiency can be completely overcome by a diet of the high lysine corn. In 1969, production in Colombia amounted to about 400 tons. Other Latin American countries are following the lead of Colombia. In India, a new surge of interest in high yielding corn is accompanied by the work of breeders introducing the opaque-2 gene into hybrids that are suited to their agricultural conditions.

The effects of these three products of the joint research of geneticists and agronomists are just beginning to be felt. Given a wise and beneficial development of the economy in the developing nations of the tropics, plus more equitable distribution of land ownership and wealth, they should go a long way toward relieving the stress of hunger and malnutrition in impoverished areas.

In accepting the 1970 Peace Prize, Dr. Norman E. Borlaug of the Rockefeller Foundation said that the present "Green Revolution" can buy only 20 more years for a world faced with overpopulation. He stressed the urgency of worldwide population control. Most nutritionists and agricultural economists also agree that worldwide starvation has been delayed by the increased carbohydrate supply resulting from the "Green Revolution." Even so, if the world's people are to be reasonably well fed two other important areas of food production and human nutrition need increased attention. Protein and vegetable oil supplies must increase concomitantly with the increased carbohydrate supplies. Fortunately, U.S. oil-crop breeders and farm industry have provided some interim answers for immediate protein and vege-

table oil needs until worldwide food-production programs
can be more broadly mobilized on the scale of the wheat,
rice, and corn programs.

Growth of the U.S. soybean industry in the last 30
years has been without parallel in the history of world
agriculture. By 1970 soybeans ranked third in the U.S.
acreage and second in value of all cash crops (excluding
hay). The wide adaptation of soybeans was made possible through the highly successful efforts of plant breeders in perfecting varieties adapted for local use almost
everywhere in the United States. Agronomists and allied
scientists have been working as members of the research
team providing a full spectrum of research information
necessary for maximizing the productivity of the improved soybean varieties.

Today, of every five bushels of soybeans we grow,
two bushels, or their equivalent in oil or meal, are ex-
ported to oilseed-deficient nations. In less than 25 years
the United States has become the world's largest exporter
of fats and oils, with soybeans ranking at the top in dollars earned among U.S. agricultural exports. The two
basic products of soybeans are oil and protein meal. Each
60-pound bushel of soybeans produces 11 pounds of
oil which, in turn, can be manufactured into shortening,
margarine, mayonnaise, salad oil, and other foods. Some
soybean oil is used for industrial purposes, in paint,
varnish, linoleum, rubberized coatings, and so on. Each
60 pounds of soybeans also produces 47 pounds of protein meal. The increasingly efficient U.S. production
of milk, butter, eggs, poultry, beef, pork, and lamb has
as its base the protein available from the large supply
of soybean meal. High quality protein soybean meal
for human consumption is a rapidly growing industry
and source of protein for people everywhere.

Recently research has been focused on safflower as a
commercial oil crop. As a product of cooperative plant
breeding efforts, lines have been developed to meet
special industrial and human food needs. Especially
selected for their high oil content, improved varieties
were grown on nearly 350,000 U.S. acres in 1970. Much
of the current interest in safflower oil as a food in the

world stems from certain medical reports that linoleic acid, from the safflower, may have some therapeutic value in preventing arteriosclerosis, a circulatory system disease in humans.

Another new star recently added to the oil-crop plant-breeding constellation is high oil-content dwarf sunflowers. Plant breeders have recently produced varieties yielding about 50% oil. Sunflower oil is comparable in quality and usage to safflower oil. The meal remaining after oil extraction is a high protein supplement for livestock, containing 40–50% protein and 10–12% fiber. Sunflower meal is unusually high in vitamins, such as thiamin and niacin, as well as minerals, and exceeds other common oilseed meals in carotene and calcium. Interest in commercial production of sunflower oilseed is mounting in the United States in areas where soybeans or safflower are not especially well adapted.

Most dramatic has been the effect on food production by the increase in the U.S. use of fertilizers and other chemicals. Research findings on how to use fertilizers more efficiently is well illustrated by the current rice crop fertilization technology. Workers in the United States and at the International Rice Research Institute in the Philippines and elsewhere have clearly demonstrated that varieties of improved stiff stemmed, non-lodging, short statured rice favorably respond to very high levels of added plant nutrients. Yields of over 10,000 kg/hectare from a single rice crop have been recorded under field conditions. This is nearly four times the yield potential of the old line lodging varieties. Purchases of fertilizers for all crops increased nationally from $868 million in 1950 to $1,771 million in 1966 and pesticides from $179 million to $619 million during the same period, enabling increased population.

Regulation of plant growth and behavior, and weed control accomplished through use of agricultural chemicals, improved plant nutrition through the use of fertilizers, and regulation of insects and pathological pests have been of great benefit to the public, both from the standpoint of information development and application, and from the standpoint of sensitizing the public to the natural and man-made health hazards in their environment. Scientists have had a twofold role in this development. On the one hand they have sought specific answers to problems of plant growth and development, plant disease, and insect problems; at the same time they have opened up new areas of research and knowledge that have had direct and indirect beneficial effects for the consumer. One of the most promising examples is to be found in the field of growth regulators.

Following a lead suggested by Charles Darwin 100 years ago, a scientist working in Holland published in 1928 a paper that showed that a plant hormone, christened auxin, was responsible for the characteristic bending of plants toward light. From this discovery, and the further research in the role of plant hormones, has developed a fantastic field which has yielded information on almost all phases of plant growth and development. As a result of the knowledge that auxins would affect growth, an enormous research effort in commercial industry has developed, the purpose of which is primarily the beneficial regulation of plant behavior. The fact that auxin at low concentrations would cause bending but that a high concentration was lethal and would cause plant death led to discovery of whole new families of plant growth regulators and weed control chemicals, such as 2,4-D, a selective herbicide for broad-leaved plant control. Not only are auxin-like compounds used as weed killers at low concentrations, they are also used beneficially in fruit production. In addition, growth regulators are used to prevent sprouting of potatoes in storage, and to promote rooting in cuttings of perennials difficult to root. Another hormone, gibberellic acid, is used to induce fruit enlargement in grapes, for induction of sprouting in potatoes, for stimulation of extension growth in celery, for stimulation of flowering in seed crops, bud formation in artichokes, and for delaying ripening of citrus. Other regulators have been found or synthesized and there is every reason to believe that they also will find commercial application.

The most recent plant growth regulator candidate is the gas, ethylene, which for years had been used to ripen such fruits as honeydew melons and tomatoes but which because of ongoing basic research, is now understood to be a natural plant hormone that controls many other phases of plant growth and development. It is being used commercially in a specially prepared liquid form, a discovery that emerged from research by organic chemists who had no direct interest in the biological application of ethylene.

The facts are equally dramatic and important regarding the impact of the development of pesticides to control insects and diseases that attack plants. Biologists have spawned a completely new approach to solving agriculture's pest-control problems in growing crops. Control of pest populations by use of pesticide has increased production of food crops—but it has also brought environmental quality concerns to the United States and the rest of the world in recent years. Chemical pesticides were introduced to provide man with an additional advantage over those insects, plant diseases, and rodents that compete with man for this food supply. In some cases, pesticides have had side effects ranging from annoyance to serious ecological consequences. Use of pesticides which pose a serious threat to animal species or are a human hazard are being curtailed and will be replaced as biologists discover better methods of pest control.

While plant breeders provide plants with greater genetic pest resistance, the entomologists are busy developing nonchemical biological pest control methods. By a dedicated interest and research in depth, scientists have heightened the public's awareness of the problems of agricultural chemicals. Without stepped-up assistance from scientists, our survival problems will indeed be intensified. (For a further discussion of controlling pest populations, see Chapter 6; the problem of human hazards from chemicals is discussed in Chapter 7).

HARVEST TO TABLE

Adequate food production in volume is only part of the recent progress. Food, as we know it today, consists of a

variety of fresh but primarily processed products. The
food item is frequently grown thousands of miles and
several days' travel away from where it is consumed. The
day is past when very many of us can produce our own
food just outside the kitchen door. In turn, we need and
want food "out of season" and we therefore need to know
how to best store fruits, vegetables, and cereals in fresh
or processed form for from several days to a year.

During the past 25 years, research has contributed to
notable advances in the postharvest physiology and required management and care of fruits and vegetables,
through improvement in refrigeration, storage, packaging, transport, supplements to refrigeration, and new
processing techniques. New varieties that better withstand shipment and storage have been developed, and
losses in transit and storage have been reduced by refrigerated and controlled atmosphere handling, control
of sprouting, and use of decay inhibitors. As a tangible
result, the American consumer has available every week
of the year a large variety of fresh fruits and vegetables
and a multitude of frozen, canned, and other preserved
products.

As America has changed from being a nation of rural
people to a nation of city dwelling nonfarmers, producers
are learning more and more about when to harvest fruits,
vegetables, and cereals and how to harvest and handle
them so they will have the characteristics that people
want when they are eaten or processed. The maturity
at which a fruit, vegetable, or cereal is harvested has
much to do with yield per acre, susceptibility to mechanical damage, its chemical composition and consequent
nutritional value, the rate of deterioration by its own
physiological processes and by the attack of microorganisms, the ultimate appearance, flavor, and acceptability
by the consumer. These considerations are as important
for commodities to be processed as for those moving
directly into fresh market channels.

Storage pest control, lowered moisture content, and
lower storage temperature have proved to be satisfactory
means of controlling deterioration after harvest of crops
such as wheat, rice, corn, oats, barley, and dry beans.

Here the biological scientist has provided fundamental information about the relationships of water content and temperature to deterioration of the fruits and vegetables after harvest. Advances in this area have been important.

Considerable additional information has been gained in regard to physiological responses of fruits and vegetables to time and temperature of holding after harvest. The greatest advances have been in connection with transit temperatures with vastly improved refrigerator rail cars and trucks. The first improvements were associated with forced circulation of air within the car. More recent advances have been the construction of much larger, better insulated cars, with mechanical refrigeration capable of temperature control ranging from below zero for frozen foods to whatever upper temperature level may be desired, depending on the commodity. For perishable fruits and vegetables sold fresh, it is now customary to control temperature between harvesting and final sale to the consumer. This is quite a change from 25 years ago. Ripening of certain fruits by the processor after harvest and before canning has also served to improve quality and lower cost.

Along with temperature control has come controlled atmosphere storage of certain commodities where it is highly advantageous to increase the storage life and therefore the period of marketability. Controlled atmosphere is simply storage at low temperature in a sealed room or building where the gases in the atmosphere can be controlled. Most controlled atmosphere storage involves reduction in atmospheric oxygen to retard the rate of respiration and enhance holding quality. At the present time in the United States about one apple in every four going into fresh consumption is held in controlled atmosphere storage. There is some commercial application of controlled atmosphere storage to pears. Spoilage by microorganisms of highly perishable fruits such as strawberries and cherries, during transit to distant markets, has been reduced by elevating the carbon dioxide content of the storage atmosphere. Technological improvements in storage rooms, rail cars, and truck vans have all been helpful in expanding the use of controlled or modified atmospheres and have contributed signifi-

cantly to improvements in controlled atmosphere storage.

The loss of water from a fruit or vegetable between
harvest and when it reaches the consumer or processor
is a major cause of deterioration. Research has shown
that this condition can be controlled by keeping atmospheric moisture high by a variety of techniques. Coupled
with cold storage, this problem of water loss is easily
eliminated.

Deterioration of fruits and vegetables by diseases is
responsible for excessive losses when the commodity has
suffered mechanical damage or is handled improperly.
Control of decay is essential for maintaining the edibility
of the product. For example, in the storage of grapes it is
essential that they receive periodic fumigation for the
control of decay organisms. In recent years, the use of
irradiation for the control of microorganisms has received a considerable amount of attention. However, only
in strawberries have decay organisms been controlled
without severe damage to the commodity. Research in
the area of controlling decay organisms is receiving much
attention and it appears likely that new approaches may
result in additional methods of solving such problems.

FOOD FROM ANIMALS

Man uses many crop plants indirectly for the production of livestock. Poultry, beef, pork, lamb, and their
products are the major sources of protein in our Western
society. Availability of adequate supplies of poultry and
meat is taken for granted by most Americans, but this
abundance of animal protein is another triumph of the
application of biological research to man's benefit. Research on improving livestock is not limited to this country; throughout the world scientists have applied basic
knowledge to production of animal protein for human
consumption.

There have been many significant developments on
forage plants for animal feeds. In New Zealand and Australia the geneticists, biologists, and plant nutritionists

have been able to create varieties of grass by adding minor elements to low productivity soil and have turned these lands into areas rich in feed for sheep and cattle. Their observations have been useful in other areas of the world and in the United States. In Puerto Rico and California agricultural technology has converted unproductive land into land economically feasible for cattle and milk production. In other areas of the world biologists are showing ways to eliminate undesirable shrubs and weeds, and are introducing varieties of fine quality range grass for meat animals. In addition, they have shown the way to make barren land highly productive. These are dramatic developments. While such developments tend to go unnoticed, they have had a tremendous impact on man's food supply.

The great impact that improved animal nutrition has exerted since World War II is reflected in the degree of improvement in the efficiency with which human food is produced from farm animals. The proportion of dietary energy ingested as chicken, pork, beef, and their products in 1970 was 25–40% greater than in 1945. Although a part of the increased rate is attributable to improved genetic composition, disease control, and husbandry, probably 50–65% of the increase has resulted from developments in animal nutrition. In 1970, there were only one-half as many dairy cows in the United States as there were in 1945, yet the total milk output was approximately the same as that in 1945. Numerous studies have revealed that no more than 20–30% of the increase in milk yield is associated with improved genetic merit, with the remaining coming from the improvement of feeding and environmental factors. This results in lower cost of milk and milk products to the consumer.

As noted in the chapter on basic biomedical sciences, studies of cell metabolism have had far-reaching effects. Investigations on intermediary cellular metabolism have proved invaluable to enhanced production of animal protein. Understanding of how cells synthesize proteins, fats, and carbohydrates has led to modification in dietary formulation used in raising livestock. Improved types of animals, raised on nutrients that supply all needed substances in the proper proportions, yield more useful protein more quickly at reduced cost.

As a result of these advances, model animal systems have been developed that are useful tools in study of the basic nutritional pathways in man. Investigations on animals have shown that the body cells, including those of adipose tissue, are dynamic. In certain animals, such as the rat and pig, adipose tissue is the major site of fat synthesis. In other animals, such as fowls, the liver is the main synthesizing organ. In addition, research had demonstrated that as the dietary fat level is increased, the net efficiency with which energy is utilized for body gain in rats, chickens, and sheep increases, but the energy cost of maintenance remains constant. Similarly, of major chemical components of the animal body, fat is the most variable, although the amount of fat in the body of rats and chickens and in bovine milk can be manipulated by dietary means. The amounts of body fat in growing pigs, sheep, and cattle is fairly rigidly associated with body mass irrespective of the nutritional treatments that have been imposed to date. The newborn or very young animal, whose growth occurs as a result of increasing cell numbers rather than cell size, is highly vulnerable to environmental influences. The behavioral changes these effect also may be important in affecting later growth. Studies of small animals and livestock have shown that several treatments of a physical nature have been found to influence the efficiency of energy utilization; for example, in ruminants, the same diet fed in small particles effects a greater storage of energy than when the diet is of a larger particle size.

In addition, related work has resulted in a more highly detailed characterization of the biochemical and pathological lesions associated with nutrient deficiencies and excesses than existed prior to 1945. A great number of interrelationships among the nutrients has been discovered. With improved knowledge of the nutritional characteristics of more food and continual reevaluations of the minimal requirements for nutrients, it became possible in 1970 to prescribe more effective diets for the support of given body functions of specific animals under various environmental conditions than was possible previously. Diets now can be formulated by computer programming.

The 30-year period preceding 1945 was an era marked by the discovery of nutrients, establishing the dietary essentiality of certain major nutrients, characterizing nutrient deficiency syndromes, evaluating food sources of nutrients, and determining preliminary minimal dietary requirements. Since 1945 the dietary essentiality of very few nutrients (for example, vitamin B_{12}, selenium, and molybdenum) has been established.

The discovery of vitamin B_{12} is of interest because it implicated other nutritional discoveries. It had been known that minute quantities of cobalt per day represents the difference between life and death in ruminants. By 1945, it was recognized that the growth response of rats, pigs, and chickens, and the hatchability of eggs produced by all-plant diets could be improved by the addition of cow manure, fish meal, or liver preparations, and the effect was not attributable to the improvement of the amino acid assortment. It had been known that liver extracts have antipernicious anemia activity. The discovery of vitamin B_{12} in 1948 was significant for it represented the complementary efforts of people in at least five distinct disciplines; and it emphasized that a nutrient may be metabolically essential for all animals yet it may not be a dietary essential for certain others. That is, ruminants, by virtue of bacterial synthesis in one of the stomachs (reticulorumen), do not require in their diet preformed amino acids, vitamin K, or the B-complex vitamins. But for ruminants cobalt is a dietary essential. Although vitamin B_{12} is essential for normal metabolism, it is not required in the ruminant diet because it is synthesized by microorganisms in the stomach in sufficient quantities for the host if adequate cobalt is present in the diet.

Subsequently it was observed that certain by-products of fermentation, when added to the diets of animals, produced a growth response. Eventually, in the 1950's, it was resolved that a part of the growth response was attributable to the small amounts of antibiotics (as well as of vitamin B_{12}) contaminating the by-products. This

observation led to the supplementation of the diets of meat-producing animals with a great variety of non-nutrient growth promotants, including antibiotics, arsenicals, copper sulfate, and various hormones.

Many other advances were made after 1945, although in some cases the work was initiated prior to 1945. The composition of the soil was associated with the nutritive value of plants grown in it, and as a consequence, the geographic distribution of many nutrient defects has been identified. Parts of some food-producing plants or other plants growing among food sources (e.g., in grazing land) have been found to be toxic. In many instances, the toxic principle has been identified and antidotes or circumventory management schemes have been devised.

In a number of animals species, it was demonstrated that nutrition that brings rapid growth does not necessarily bring high productivity during later life or a long life-span. But presumptive evidence was gained to indicate that nutritional deprivation, especially that of protein or thiamin, during infancy reduces exploratory behavior, "intelligence" or learning ability, memory, and brain development in rats and pigs. The implications of these to human nutrition and development are under study.

Studies in animal nutrition have complemented basic investigations on animal physiology. During the past 20 years physiological discoveries have greatly changed the economics of livestock production by the use of diethylstilbestrol, artificial insemination, and manipulation of environment.

Diethylstilbestrol is a synthetic hormone used in accelerating the rate of weight gain in cattle and sheep. When used properly, it results in high quality, safe meat. This hormone will increase the rate of gain about 15% and in a 120-day feeding period an implanted steer will put on about 50 lb. more weight than an untreated animal. This weight is put on without any appreciable increase in feed costs because the hormone stimulates protein synthesis and depresses fat deposition slightly. At current prices this means $15.00 extra income, whereas the cost of implanting (hormone and labor) would be less than a dollar. About 25 million cattle go through the feedyards annually in the U.S.; 95 % are treated with the hormone.

Studies on reproductive physiology have led to widespread use of artificial insemination in domestic animals. From 1945 to 1965, the number of cows bred artificially increased from less than 500,000 to almost 8 million. Presently, about 50 % of all dairy cows are artificially inseminated. The frozen semen technique means that cows in any part of the world may be inseminated with semen of a given bull and that the semen of a bull can be stored and used after he dies. Not only does artificial insemination obviate the necessity of every small dairyman maintaining a bull, but, more importantly, it greatly facilitates the dissemination of superior germ plasm. It has been an important factor, together with progeny testing and higher quality, more concentrated foods in the dramatic increase in milk yield per cow. Artificial insemination has practically saved the turkey industry because of the breeding problems encountered in normal mating of the more nutritious types of broad-breasted turkeys.

An important discovery of environmental physiology has been the relationship of lighting schedule to egg production. Much has also been learned on how the environment can be improved for cattle in hot arid areas. This explains why the cattle feeding industry has expanded greatly in the Southwestern portion of the United States.

Other discoveries which have had important impacts on the efficiency of livestock production include knowledge of the neurohormonal mechanism involved in controlling "let-down" of milk which has resulted in improved milking practices and in improvements of the milking machine. Our knowledge on how to depress foaming in the ruminant digestive tract has made it possible to make greater use of green legumes for cattle feed, a practice that previously led to bloating and death of animals.

The science of genetics has contributed to improved

production of livestock. Animal geneticists discovered
that some traits such as coat color may depend on a single
pair of genes. In these instances they have developed simple progeny testing procedures for determining the presence of recessive genes for specific traits in any given
sire. The testing has proved useful in fixing coat color
(red versus black in Angus cattle) and in purging deleterious recessive genes such as those for mule foot, hair-
lessness, or dwarfism in cattle.

Most traits of economic importance depend on many
pairs of genes. Classical Mendelian genetics cannot be
used in improving such traits. Rather one must use concepts of population genetics. It has been found that about
40 % of the variation between growth rates of steers of a
given strain depends on inheritance. This trait has a
heritability of 40 % indicating that great progress can be
made for this trait by selection by superior breeding animals. The efficiency with which the digestive system of
the animal converts feed to muscle is highly correlated
with the rate of weight gain. Thus selection of animals
with more rapid weight gain results in improvement in
the feed efficiency, or conversion of animal feed to meat
for human consumption. This finding eliminated the
need to carry out difficult and expensive feed efficiency
testing of breeding animals.

On the other hand, litter size in pigs has a heritability
estimate of about 10–15 % indicating that improvement
from selection would be slow. Geneticists have found,
however, that traits of lower heritability respond to crossing of appropriate lines or breeds resulting in "hybrid
vigor." Selection within the lines or breeds results in a
slow but permanent improvement in such traits while
crossbreeding between the improved lines or breeds
yields a temporary but immediate improvement.

The application of the above findings has revolutionized the production of eggs and poultry meat. Unique
lines of chickens are developed, through selection and
inbreeding, which are superior in some traits; then matings are made among lines. For example, one line may
have superior egg production and egg quality but inferior egg size. A second may have average performance

in egg traits but have superior livability and laying persistence. A third may be average for these traits but have superior egg size. The crossing of these three lines could result in a superior crossbred pullet. Over the past 25 years, egg production has increased in the United States from approximately 160 to 240 eggs/hen per year. Similar advances have been achieved in meat production by utilizing specialized strains and crosses to produce rapidly growing, highly efficient "broilers."

The development of crossbreeding systems for swine, sheep, and beef, using primarily breed crosses, has resulted in increases of 20–30% in total production efficiency. A significant portion of this increase has resulted from the discovery that crossbred mothers have superior reproductive efficiency and mothering ability. The same genetic approaches are now being applied in fish production.

The discovery of accurate methods of estimating the genetic value of breeding animals has enhanced the use of performance testing and progeny testing systems of selection. The Dairy Herd Improvement Association (DHIA) has developed computerized cow record and bull progeny testing systems based on traits important to genetic improvement and efficient herd management. During the past 10 years the average milk production per year of all dairy cows has risen to almost 12,000 pounds, an increase of over 2,000 pounds. Cows on DHIA test average over 14,000 pounds with a single record of over 40,000 pounds having been recorded. The development of artificial insemination has advanced the use of progeny tested, superior dairy bulls and the use of these superior animals as semen donors is one of the basic reasons for the tremendous increase in production per cow. This is an excellent example of a multidisciplinary approach to improving production efficiency: genetics for determining the prepotence of a sire, and physiology in extending the use of proven sires through artificial insemination. Discoveries in nutrition and management have also contributed to the success of genetic advances through providing an environment in which the potential of the genetically improved stock can be realized.

As noted previously, research and control of many human diseases has been inhibited because of necessary constraints on experimentation with human subjects. Animal models of human disease have fostered progress on many problems including not only human afflictions but also animal health. Veterinary medicine is to livestock as medicine is to man. Untold millions of dollars worth of food-producing animals have been saved for human use by the development of vaccines for the following livestock diseases: hog cholera (live virus adapted to swine cells), rhinotracheitis of cattle (virus modified by growth in tissue cultures of bovine kidney cells), bluetongue disease in sheep (virus modified in chicken embryos), Newcastle disease of chickens (virus grown in chicken embryos).

Veterinary medicine has developed a vast knowledge of blood group systems in several animal species. Application of this research is most useful in cattle where blood group data can be used to determine the authenticity of pedigrees in purebred animals. This information has been applied in artificial insemination programs where the genetic quality of the semen must be maintained. The research on artificial breeding and the development of that program have had outstanding results for improving the quality and volume of meat and dairy foods for the human population.

Agriculture and livestock production have benefited greatly from programs of animal disease eradication. Two major examples are bovine tuberculosis and bovine brucellosis. The responsible bacteria cause serious diseases in man that are contracted from the animals or animal products. Thus, the concept of disease eradication benefits the animal population and concerned agricultural industry. In addition, it removes the disease agent from the environment, which has an obvious result on human health. The concept of eradication has been successfully applied to other diseases of animals such as vesicular exanthema of swine. Another great problem that has plagued the animal production industries of the

Southeastern states for years is screwworm infestation. The research which has led to the control of this serious menace is reviewed in the Chapter on population biology.

PRESERVING AND DISTRIBUTING FOOD FROM ANIMALS

Production of animal protein is only one portion of our vast food industry. Preservation and distribution of wholesome meat products to the consumer is another area. Major advances in meat science and technology include the understanding of basic physiological and chemical properties of animal muscle and meat constituents. Physical methods of measuring the tenderness of both uncooked and cooked meats have been developed. These give a high predictability of the tenderness observed in the cooked meats by trained taste panels, and have been of value to selection of animals with improved tenderness in breeding programs and in devising the best methods for heat processing of various meat cuts, primarily beef.

Major studies on muscle protein focus on the reactivity and stability of muscle pigments (derivatives of myoglobin). These studies have been particularly significant in understanding the chemistry of the curing reaction of meats and the role of these reactions in assuring uniformity and stability of the cured meat color and in avoiding undesirable side reactions that would lead to discolorations in the cured meats. In addition, studies with fresh meats on the stability of the meat pigments have assisted in developing improved packaging methods and techniques of storage.

Studies with other muscle protein have assisted in providing understanding of the water binding and emulsifying properties of meats, which are of major importance in connection with the desired functional, keeping, and other properties of processed meats and meat products such as sausages and frankfurters.

PROCESSING AND PRESERVING FOODS

The most dramatic change in processed foods during the past 25 years has been increased convenience, which enables the preparation of foods rapidly and effectively with a minimum of effort. Hours of food preparation once done in the home are now done elsewhere. To accomplish this an enormous amount of research involving chemistry, biochemistry, physical processes, and especially the biological processes has been needed. Advances in packaging, too, have been important in convenience foods. Use of pre-prepared and packaged foods have made time previously used for food preparation available for other human activities.

Processed foods with desirable characteristics have consumer acceptability. Not only have we developed an understanding of what acceptability is, but also the psychological, chemical, biochemical, and physical factors involved. As a result of extensive research during the past 25 years, the consumer has been able to obtain in his foods a higher degree of acceptability and stability than ever before.

By stability we mean the retention of the quality of the food that it had at the time of processing including color, vitamin content, quality of proteins, prevention of rancidity and spoilage. Shelf life of foods has been increased and the consumer can be certain that when he purchases foods they will have retained much of the original quality. American foods are without doubt the safest in the world—this is in spite of what we hear about the possible harm from pesticides and additives. The scare stories in this area are anything but true; over 93% of the illnesses associated with foods are caused by microbial intoxications rather than chemicals. Furthermore, of the remaining 7%, very little, if any, can be attributed to food additives or pesticides.

Great advances have been made in the retention of the nutritive values in our processed foods. As a matter of fact, it has been shown that if fresh products are handled improperly before they reach the table, canned and frozen foods may very well be richer in vitamins than the fresh. We have learned a great deal about the retention of vitamin and nutritive quality of proteins; this technology is applied now by processors during the packing and storage operations.

Enormous advances have been made in packaging as a

result of research. Without new and improved packaging it would not be possible to have the variety and quality of foods we have today. As a matter of fact, many of the products now available are dependent entirely on advances in packaging. In Japan, for example, fish sausage containing processed fish, pork fat, smoked flavor, certain spices, color, and a preservative would not have been possible without the development of suitable packaging. Today this is an extremely important product of Japan, and in fact, has actually raised the average protein intake.

Without the extensive amount of biological, chemical, biochemical, microbiological, and physical data collected relating to the production, preservation, storage, and distribution of foods, it would not have been possible to improve the food supply as we have. Still, many areas of the world are lacking in food. Much of this relates to agricultural production, but some relates to the elimination of waste in countries where it is needed most. This has been and can be done to a greater extent by the use of available technology.

In the development of new varieties of crop plants, modified cultural practices, and improved types of livestock, changes may occur that influence the canned, frozen, or dried food products. Often there is a correlation between quality, yield, and cost. Food science has been an area that has overlapped with agriculture and when this was realized, great advances were made in handling, processing, and distribution of agricultural products.

As a result it has been possible to maintain a high quality food supply at a low cost. This is of direct importance to the consumer who has therefore benefited from the point of cost and quality of the foods available to him.

CONCLUSION

It is now recognized that intensification of agriculture requires the combined efforts of science, government, private industry, producers, and public support. Science provides the needed technology but that technology must be tailored for the diverse conditions of soil,

climate, and food preferences of each region and
nation. Scientists can and do provide the opportunities
for new technological developments, but whether or not
these new techniques will be used beneficially becomes
largely the responsibility of other individuals and organizations. The scientists, both basic and applied, are linked
in a system to meet the needs of man. There is no doubt,
however, that scientists are an essential link in a chain of
activities which add up to man's social as well as economic development.

Although some economists and sociologists now are
warning that temporary local surpluses of rice and wheat
resulting from the "Green Revolution" and new advances
in the technology of food processing will produce major
social, political, and economic upheaval in the developing countries, it still is crystal clear that from a worldwide
standpoint major increases in food, feed grains, and
poultry and livestock products will be needed in each
decade ahead. The world's population continues to grow
about 3 % a year in many of the most populated developing countries. The challenge is not whether to expand
or reject the "Green Revolution," but how to use it in a
way that will provide the greatest possible advantage for
all people. Adequately fed, clothed, and housed people
should and must find the time and resources needed to
resolve social and economic problems, thereby providing
the foundation needed for a peaceful world. Scientists
in the biological, physical, and social fields must continue
to provide new discoveries out of which the foundation of
a satisfying life for mankind may be expanded and
strengthened in the years ahead.

SELECTED ADDITIONAL READING

1. BROWN, L. Seeds of Change: The Green Revolution and Developments
   in the 1970's. New York: Praeger, 1971, p. 122.
2. BROWN, L. R. Nobel Peace Prize: Developer of High-Yield
   Wheat Receives Award. Science 170: 518–519, 1970.
3. CHANDLER, R. F. The International Rice Research Institute
   Annual Report 1968. P. O. Box 583, Manila, Philippines.
4. Crop Production. 1970 Annual Summary USDA Statistical Reporting Service, Crop Reporting Branch. CrPr 2-1 (70). Dec.
   18, 1970.

5. GRILICHES, Z. *Science* 132: 275, 1960.
6. HARRAR, J. G. The *President's Review and Annual Report 1969.* New York: The Rockefeller Foundation, 1969.
7. *Major Uses of Land and Water in the U. S. with Special Reference to Agriculture.* Summary for 1964. USDA, ERS, Agr. Econ. Report 149. November 1968.
8. MARTIN, J. H. AND W. M. LEONARD. *Principles of Field Crop Production.* New York: Macmillan, p. 322.
9. MIKKELSEN, D. S., AND N. S. EVATT. *Soils and Fertilizers. Rice in the U. S.: Varieties and Production.* Agr. Handbook no. 289. ARS, USDA, Superintendent of Documents, U. S. Govt. Printing Office, Washington, D. C. 20402. 1966, p. 65–73.
10. REITZ, L. P. New wheats and social progress. *Science* 169: 952–955, 1970.
11. *Rice Research and Training in the 70's.* The International Rice Research Institute, P. O. Box 583, Manila, Philippines, 1969.
12. STAKEMAN, E. C., R. BRADFIELD AND P. C. MANGELSDORF. *Campaigns Against Hunger.* Cambridge: The Belknap Press of Harvard University Press, 1957, p. 328.
13. STAUB, W. J., AND M. G. BLASE. Genetic technology and agricultural development. *Science* 173: 119–123, 1971.
14. *The World Food Problem.* Report of the Panel on the World Food Supply, vol. II. Washington, D. C.: Govt. Printing Office, 1967.
15. *World Food–Population Levels.* Report to the President. USDA in coordination with the U. S. Dept. of State, Agency for International Development. April 9, 1970, p. 1–4, 7, 8.
16. WORTMAN, S. Progress Report: Toward the Conquest of Hunger 1965–1966. The Rockfeller Foundation Program in Agricultural Sciences. New York: The Rockfeller Foundation, 1966.

Chapter VI

Population biology

FREDERIK B. BANG, M.D.

THE GREAT ADVANCES during the past 30 years in medical and dental sciences, in marine and aquatic sciences, and in food production and technology have had a part in generating environmental factors that increasingly threaten human welfare. Population biology is concerned with the factors that affect both the quantity and quality of the human populations, but this branch of biology also concerns itself with populations of other organisms. Man eventually relates his population levels to those of other animal and plant communities, and he regulates his environment in a way that will benefit both himself and his total environment.

Man is concerned primarily, but not exclusively, with the quantity and quality of the total human population and has begun to move toward limiting the former and controlling the latter. By successfully decreasing pest populations and reducing certain diseases through control of populations of disease-transmitting organisms, man has attained a healthier phase of his own population. But no growth process goes on forever unimpeded, whether it be a population of cells in an animal or plant body, a population of animals or plants, or the human race. Restraining factors increase with the increasing density of the population and will slow population growth either gradually and smoothly, or suddenly as in catastrophic events. In one way or another, man's population will be limited; the question is how and when. We now have newer, simple and safe methods for the

215

control of human reproduction, and these methods are continuing to be improved. The use to which man will put this newer scientific knowledge on control of population quantity has enormous implications in man's future.

Since ancient times man has been concerned also with the quality of human populations. Many philosophical discussions have probed how to "improve" the human race, both as individuals and in groups. Students of human genetics have delineated many genetic defects and have worked out mechanisms of inheritance, even at the molecular level. These developments have led to a new service to mankind: genetic counselling. Thus far, this is of benefit to individuals and not to population groups. The regulation of the genetic makeup of human populations cannot at present be predicted, even if desirable.

It is not feasible to do research on a total population in this world at a single time, whether it is the human population or those of an animal or a plant. Consequently, population research consists of projects focused on smaller groups of organisms. It is possible to study some limited groups in localized environments wherein the interrelationships of individuals to other members of the population and of the organisms to their physical environment are known. Such an "ecosystem" then can be viewed as a model of a larger ecological system.

Research on several aspects of population biology has led to development of techniques useful in controlling harmful plant and animal populations. Of benefit to man have been the control of insects that transmit disease organisms, of the screwworm that is a major pest of livestock and large game animals, of agricultural insect pests and weeds, and of animal pests such as rats and rabbits. Population research is also concerned with the preservation of populations of certain animals and plants beneficial to or even necessary to the preservation of human health and welfare.

POPULATION POTENTIAL AND CARRYING CAPACITY

Although population research does concern itself with the change in numbers of an animal or plant species within a circumscribed area during a period of time, it does more. During the past 30 years tremendous advances have been made in the mathematical analysis and representation of processes that go on in populations, in identification of the forces involved and in the formation of laws by which they operate. In addition, population research has developed two useful concepts: *1)* "population potential," which provides insight into densities of human populations in different regions that are consistent with effective economic conditions and communications, as well as insight into the effects of these regional differences on human health; *2)* "carrying capacity," the capacity for a particular environment to provide resources for a population of optimum size. The second concept has become the cornerstone for the management of wildlife and the harvesting of excess individuals.

Studies of the environmental setting of many species have shown that each occurs over a range of habitats, some of which support more individuals than others. Analysis of these different environments has provided firm criteria that influence optimum population density: factors such as climate, physical geography and the composition of associated species. These criteria have been useful in increasing the carrying capacities of certain environments relative to certain species.

In research on experimental populations, artificially structured environments have been devised that simulate variables that occur in natural settings. Out of this type of research have come ideas for modifying organization of limited space and concepts of efficient use of limited space that have attracted the attention of architects and city planners. For example, space needs for common activities such as transportation to and from segments of an area can be computed from models. Similarly, studies of carrying capacity have led to knowledge of the behavioral effects of crowding. It can be anticipated that application of these principles will significantly and beneficially affect the human scene in the future.

Population researchers have developed sophisticated methods (systems analysis) of analyzing total energy flow within an ecosystem, based on the concept of "food chains." The food chain concept relates to the capture of energy from the sun by photosynthesis in green plants and the flow, so to speak, of this energy through successive species of animals that use each other for food, all the way up to the largest predator of the particular ecosystem. These insights provide a basis for defining the ultimate limit to the number of living organisms, including man, that can exist on this earth. In this connection, there is now an indication that fish will provide an increasingly larger percentage of the protein requirements of man. As noted in Chapter 7, the study of fish populations has formed a basis for more effective exploitation of this resource. But further knowledge of food chains in the sea is needed to prevent damage by environmental contamination which, as pointed out previously (see Chapter 3), is already beginning.

An outstanding example of the successful application of basic research findings to a population problem is the management of the Alaskan fur seal which has been brought back from a point of near extinction. Fur seal populations of the world were largely destroyed by ruthless harvesting. However, international agreements of 1911, 1957, and 1963 concerning the Pribilof seals have largely forbidden wasteful oceanic sealing in return for international distribution of 30% of the harvest from the carefully controlled breeding grounds. Under this protection, the Pribilof seal population increased from a quarter of a million in 1912 to about 2 million in the 1950's; since then it has been deliberately reduced to just less than 1.5 million in 1967. Since 1920 the seal population yielded a rich harvest of 60,000 prime skins a year. Management of the seal herds has increasingly depended on long-term research on all accessible aspects of seal demography: the social organization of the rookeries and outlying groups, disease rates on the rookeries, and feeding habits over wide areas of the oceans.

About 700,000 seal pups have been marked with numbered metal tags. Currently 3,000 1- to 3-year-old seals are tagged each year, and 10,000 pups are temporarily marked by shearing the fur. The number of

pups born each year can be estimated by combining
data on recovered tagged animals, the ages of the tagged
seals that are harvested (determined from growth rings
on the teeth) and pregnancy rates. Subsequent calculations on number of pups obtained from these data can
be rechecked against estimates from the capture of
marked pups on the beaches, from counting dead pups,
and by complete counts of pups in small areas. When
the Pribilof population was small, less than 5% of the
pups died; in the 1950's up to 20% died on land, and
there were heavy and highly variable death rates of
young at sea, apparently from food shortage. After 1955
the harvest of 3- and 4-year-old males was variable and
unpredictable, ranging from 96,000 to 30,000 a year.

After 1955 females were harvested, and recently larger
males, too, to reduce the population to a level that
produces steady harvest. The policy of taking females
was against popular opinion, but its value has been
proved by results.

Mathematical models of fur seal population dynamics
and harvesting are not yet complete. However, management based on present understanding of the harvesting
dynamics is an ideal example of how a wild population
resource should be used so that the wild population
flourishes while people reap a continuing, abundant
harvest.

Much of the early research on human population
took a statistical point of view. Man was seen as just a
cog in a gigantic social machine and the average impact
of environmental factors was the primary consideration.
This approach led to studies of the variability among
members of a population which indicated that individual
human welfare was influenced by genetic, developmental, and constitutional factors. This has led to a
shift in the point of view of much of population research
from the statistical to a focus on the worth of the individual member of the population and to the population
as a milieu for individual growth and development.
Along with this shift, there has been an increasing
emphasis on the behavior of the individual, especially
with respect to aggressive and violent behavior. It must

be kept in mind that population biology includes the behavior of populations which, in turn, are dependent on the behavior of individuals that make up the population.

Research has shown that most vertebrates form dominance hierarchies in which initial overt aggression becomes sublimated into a system of mild threats and coordinated submissions. Once such social roles become established, the amount of aggression and attendant physiological stress become much reduced. Furthermore, any population over a period of time works toward a segmentation into a number of groups of an optimum size for the species. Within each group the stabilized hierarchies provide maximum opportunity for individual development. However, when the population gets too large, individuals find no opportunity for their integration into an established group. Their attempts to become socially involved are repulsed. Such rejected individuals become exceedingly withdrawn both physically and psychologically and soon lose the capacity for social involvement. Overpopulation, however, leads to a much more insidious alteration of behavior than merely the development of violence.

Research with rodents has shown that in overcrowded rodent populations young animals subjected to the high rate of contact, unavoidable in dense population, never develop the capacities for expressing most of those behavior patterns essential for the survival of the species. Such individuals never develop the capacities for aggression, integration into a social group, courtship, copulation, or maternal behavior. As a consequence of this noninvolvement, they are physiologically unstressed but, so far as behavior is concerned, they hardly justify being called a mouse or rat, as the case may be. Though physically adult, behaviorally they remain at the juvenile stage of development. These and similar studies may provide clues as to some possible consequences of overpopulation for man —the inability to attain the genetic potential of becoming a sophisticated, cultural being.

The concept of "stress" has proved to be a pivotal

one for uniting research efforts in physiology, psychology, and population dynamics. Many measures of disturbance of normal physiology may now be utilized to assay the degree of crowding, stability of social relations within a population, and adequacy of the structure of the environment.

Growth and development, both physical and behavioral, have been shown to be effective measures of population dynamics. Even when the quantity and quality of food is optimum, overpopulation or social disorganization among its members leads to inadequate utilization of the food and to the stunting of physical growth. An example of abnormality produced by normal dietary constituents is represented by vitamin A. An intake of only four times the normal level in slightly overcrowded rats produces individuals with inadequately developed behavioral patterns somewhat similar to those that arise with a normal diet associated with excessive overpopulation densities.

The field of population genetics has shown clearly that the basic unit within which evolutionary change becomes expressed is the population; that is, the whole gene pool of a population and its successful development. Promotion of better quality of life for man may only become fully realized by a betterment of the quality of the heredity genetic pool from which future individuals will be drawn. Furthermore, research has revealed the advantages of genetic diversity. Populations whose members differ from each other in their heredity exhibit superior capacities to adjust to altered environmental conditions and to survive through successive generations. Similarly, on the individual level, crossbreeding or avoidance of inbreeding usually enhances the inherent capacity of the individual to cope with exigencies encountered in his particular environment. However, enhanced genetic potential is of little value unless it can be coupled with an environment increasingly able to nurture the undeveloped creativity within the genetic pool.

Study of the evolutionary process out of which man has emerged reveals that those forms that have survived up to any particular point in time have in general been those which *1)* are more perceptive as to conditions and changes of their environment and *2)* are more capable of initiating adjustments to changed circumstances. Though these two aspects of changing life are closely linked, the enhancement of awareness stands cut as the central process in evolution, providing greater chances of survival to those populations whose members in general have greater capacities for awareness than do members of other populations. This process has continued as the cultural evolution of man has become superimposed on genetic evolution.

Research has shown that no matter how adequate the heredity of an individual may be or how adequate the total genetic pool of the population of associates in which he is immersed, his fate will be drastically influenced by the circumstances surrounding his development. For this reason, the fields of developmental psychology, ecology, and population research have become more and more fused as they jointly seek more understanding of how to enhance the fulfillment of both genetic and learned capacities. As research so often does, the initial approach encompassing the largest body of effort has emphasized the abnormal. It focused on the effects of physical and social isolation; that is, it has focused on a population reduced to one individual. It has been found that long-term isolation reduces the individual's capacity to learn to adjust emotionally to strange situations and to become integrated into a group. With this base line, research has become focused on development with increasingly larger and more complex groups to the opposite extreme of pathological crowding. As a result, there has developed a body of knowledge which says that there must be an optimum rate of exposure to new situations, including other individuals, if an individual is to achieve the fullest development of his innate and acquired capacities.

Population research has revealed that the effective functioning of a population requires its fragmentation into smaller aggregates. Such aggregates enhance escape from predators, construction of communal shelters, acquisition of food, the integration of young into adult life, and the gratification that comes from membership in a society. Studies have shown that the social way of life is ingrained in practically all lineages of animals, including that lineage from which man has stemmed. Research has led to the important conclusion that for every species of animal there is some optimum number of contacts per day per individual, if the individual is to maximize the meaningfulness of these relations for his own life. Beyond this general conclusion there exists a huge body of knowledge relating to the strategies animals follow for maximizing meaningful social relations. These ideas, derived from basic animal research, are proving helpful to those who are attempting to grapple with comparable problems at the human level.

Every contact (even those occurring at considerable distance where smell, hearing, or seeing are involved) presents the opportunity for some meaningful communication to occur. For example, some studies on vocalization and audition of animals were focused on the mechanisms involved while, quite independently, other studies were focused on the use by animals of communications by audible sounds to weld their societies together more effectively. As such independent efforts began to overlap and unite, it became possible to describe animal societies in terms of communication networks and systems. This has led to the recognition that all populations and societies simultaneously metabolize two kinds of information: food and ideas. These insights reaffirm the adage, "Man does not live by bread alone." The conclusion has been reached that the increase in evolutionary awareness ("progress") is measured by the ratio of energy or time a species devotes to acquisition and use of "ideas" versus the acquisition and use of food.

POPULATIONS IN ECOSYSTEMS

The development of the concept of "ecosystems" has led to an overall implication for human welfare of the research on and related to the problems of populations. Out of basic research in its purest form has come the understanding that all populations of all animals and all plants are inextricably bound together in a mutually supportive life system. Other species suffer when any one species makes unusual demands on resources. And when many species suffer too long, the species that disrupted their relations among themselves will soon, in turn, suffer from deficits or abnormalities of the ecosystem thus produced. Man now is in the position of beginning to suffer from his destructive impact on his ecosystem. Most of the present public concern about population control, pollution abatement, and environmental rehabilitation could never have arisen without the huge body of basic research that has been done on ecosystems.

One of man's efforts to better his environment was his attack on populations of insects that transmit agents causing human disease. The golden age of discovery in the field of medical entomology occurred at the turn of the century, when biting insects and related arthropods were shown to be the transmitting agents of malaria, yellow fever, typhus, plague, and other diseases. In subsequent years, the basic information on the epidemiology and natural history of the arthropod-borne diseases has continued to accumulate. The malaria vectors were sorted out from the nonvector anopheline mosquitoes; yellow fever was shown to exist in the canopy of the great tropical rain forests of Africa and South America where monkeys served as reservoir hosts; and studies on the host relationships of the tick and mite vectors of Rocky Mountain spotted fever, scrub typhus, and other rickettsial diseases showed why these diseases are able to maintain themselves. Investigations on the natural history of the encephalitis viruses have revealed that many species of birds are reservoir hosts for many types of these viruses that, under certain conditions, are transmitted through mosquitoes to horses, related species, and man.

CONTROLLING PEST POPULATIONS

The basic epidemiological information permitted intelligent planning for control of these diseases by attacking the arthropod vectors at some particularly vulnerable stage in the life cycle. The great breakthrough was the discovery of DDT. A fine deposit of DDT on the inside walls of houses destroyed mosquitoes, fleas, and other disease vectors; application of DDT to military personnel and civilians prevented outbreaks of typhus during World War II. It would be difficult to exaggerate the impact of DDT on public health. Malaria, which had been a most important cause of ill health and death, especially in the tropics, responded so rapidly to this new method that in 1954 the World Health Organization elevated the objective of its worldwide campaign from control to eradication.

For several reasons eradication has not yet been achieved. According to a 1968 report by the World Health Assembly, by 1967, of the 1.7 billion people who had lived in malarious areas over 1.3 billion were enjoying the protection offered by DDT.

To quote the World Health Organization (WHO) Expert Committee on Malaria in its Thirteenth Report: "To have progressed this far and to have brought a sustained measure of well-being to a total of 953 million people, more than one-quarter of the world's population, is an international achievement without parallel in the provision of public health service."

DDT was followed by the development of other compounds for insect control. These, as well as DDT, are especially effective because of their stability and have had wide usage, not only for control of diseases, but also for the control of insect pests in forests and agriculture. It has been reported that the peak consumption of DDT in the United States was 79 million pounds in 1959. By 1967–1968 this usage had declined to 33 million pounds; nevertheless, these kinds of pesticides continue to be applied in very large amounts all over the world.

The advances in the technology of pest control have created new problems, involving insecticide resistance, environmental pollution, and hazards of accidental exposure of man and animals to poisonous substances. These environmental hazards are discussed in another chapter. There is mounting public pressure to reduce drastically, or even eliminate, the use of many or all of the new compounds. Considerable effort, therefore, is being made to find alternate methods for controlling disease carriers and pest arthropods.

The WHO, in cooperation with laboratories in several countries, has been conducting a screening program in which, so far, over 1,300 compounds have been tested as potential replacements for DDT in malaria control. Some show promise and are already in use; but for the most part these alternate materials have disadvantages in that they are too expensive, have an unpleasant odor, are too toxic to man, or lack the residual properties which are essential for malaria control.

Persistent insecticides do not break down rapidly but remain in the environment for long periods. The replacement of persistent by nonpersistent insecticides or alternate control techniques that do not depend on insecticidal chemicals are especially desirable. Compounds which will break down rapidly do not contribute so much to environmental pollution. Some insecticides, less stable than the others, have been used in ever-increasing amounts. Some insecticides that can be used safely are extracted from plants; these include pyrethrum, rotenone, nicotine, and ryanodine. Insecticide formulations and the various kinds of apparatus by which they are applied are undergoing constant modifications. One method involves low volume spraying, in which very small amounts of a highly concentrated insecticide are applied.

Sterilization can be brought about not only with irradiation, but also with a group of nuclear poisons known as chemosterilants. The advantage of these substances is that they can be applied in baits, or by other means, to field populations, and females as well as males are affected. There remains. however, a serious problem of safety in field application.

Experimentation with methods of genetic control is for the most part still at the laboratory level; it includes studies on lethal genes, behavioral modifiers, distortion of sex ratios, hybrid sterility, cytoplasmic incompatibility, and translocation semisterility. An interesting field experiment was conducted in Burma, where a

population of the common house mosquito was apparently eradicated by the release of cytoplasmically incompatible males. Field experiments on sterile males, produced by crossing of two strains of an *Anopheles* mosquito, are now under way in Africa.

Certain substances, called pheromones, are produced by insect glands and remain active even outside the insect's body. They may exert a powerful attraction to the opposite sex and they may guide the insect to its food or home, or to the proper oviposition sites. Other attractants may be found in plants, and some synthetic chemicals may also have this effect. It is appealing to think of controlling a pest species by luring one or both sexes to a trap baited with such substances. This technique was useful in eradicating infestations of the Mediterranean fruit fly in Florida in 1957 and 1962.

It has been suggested that juvenile and molting hormones of insects, which produce effects such as failure to mature or inability to enter diapause, may be exploited for control. It is possible that antimetabolites may be developed to interfere with the normal physiology of the pest species.

During World War II several plastic solvents were shown to be highly effective against mites that transmit the agent causing scrub typhus. Repellents, when properly used, also protected troops and civilians from infections with malaria and other insect-borne diseases. An improvement over the earlier repellents has been accomplished.

There is a fascinating problem related to repellency. Why are certain species and varieties of plants or animals not attacked by some insects? Is it that they exude a repellent substance or that they lack an attractant? Antifeeding compounds prevent insects from feeding on treated material, and certain ones have been used for years for mothproofing. Recently some compounds have shown promise of protecting plants and fruits against caterpillars and other chewing insects.

A growing understanding of the genetic basis of insecticide resistance is based on a return to classical genetic research on house flies, mosquitoes, and other insects. Discovery of marker genes and their linkage groupings has permitted the identification of many genes which confer resistance to various poisons. Crossbreeding and cell classification have revealed the existence of cryptic species in an *Anopheles* population and is aiding in unraveling the complexities of malarial epidemiology in Africa.

Successful introduction of predator species depends on the selection of strains genetically compatible with the new environment. Discoveries that will lead to the replacement of chemicals with such things as insect hormones and other natural substances will depend on continuing research on fundamental insect physiology. If the hazards resulting from overuse of insecticides such as DDT and other compounds are to be reduced or eliminated, it seems that the entomologists will have to utilize what they call "integrated control." This means the use of the most effective method, biological or chemical, that will bring about the desired effect. If circumstances dictate the use of an insecticide, it is applied at dosages and at times which will bring about a reasonable reduction in the pest species, but with minimum effect on predators and other nontarget species, and without leaving dangerous residues. Now it becomes obvious that one would be unaware of the most safe and effective method without an intimate knowledge of the natural history of the target species, and of the other insect and vertebrate populations with which it is associated, in other words, the entire ecology of the problem area.

Progress in the solution to problems of pest and arthropod-borne disease control, whether by a dramatic discovery of some panacea, new chemicals, exploitation of biological phenomena, or improved application, will not occur without continuing and painstaking basic and applied research.

Insect-transmitted disease has been eliminated from certain areas of the world without elimination of the transmitter (the vector). Malaria has been eliminated from the United States. This is because vectors are successful transmitters of disease only if their effective density is sufficient to maintain a population of the parasite. Thus the present effective barrier to the life of the parasite makes it unlikely that malaria will ever successfully reestablish itself in the United States. In other areas of the world, where the effective density of contact between the *Anopheles* mosquito and man has been greatly reduced by application of DDT to the inside of houses, malaria has disappeared for a time, but reintroduction into old areas has at times caused new epidemics. In many cases development and use of new insecticides has prevented effective reintroduction. Thus the continued demonstration of health in the tropics such as in the Canal Zone of Panama is dependent on continued study of the ecology of malaria, yellow fever, and various other virus diseases.

The screwworm is a major pest of livestock and large game animals, and on occasion will also attack man. The pest has largely been eliminated from the United States by the use of sterilization techniques, a unique biological control procedure developed by scientists of the U.S. Department of Agriculture. This insect, about 3 times the size of a common house fly, deposits its eggs in wounds in warm-blooded animals. The eggs hatch into larvae that feed on the animal's tissues. When mature, after 5–6 days, the larvae leave the wounds, drop to the ground, and change into the pupal stage. The adult fly emerges from the pupa after about 8 days during warm weather to start the cycle over. The life cycle requires about 3 weeks during the warmer months of the year, but may require 2 months during cool periods.

This parasite is endemic in the southwestern part of the United States, Mexico, and Central and South America. It became established in the southeastern part of the United States in 1933 where it overwintered in Florida and spread northward and eastward, each summer causing damage to livestock and game in several States. Livestock growers in the Southeast estimated losses averaging $20 million each year. In the Southwest the insect survived the winters in the southern parts of Texas, New Mexico, Arizona, and California as well as in Mexico. In the summer, it was capable of spreading northward and eastward into about a dozen states, sometimes as far north as Iowa and South Dakota. Losses due to the southwestern infestation, according to livestock estimates, amounted to about $100 million each year prior to the organization of suppressive efforts. These losses were due to deaths or damage to sheep, goats, cattle, horses, and other livestock, and particularly were due to the cost of the labor involved in the inspection and treatment of animals. Losses to wild animals, especially deer, were also often very high.

Initial research led to the development of wound dressings and animal management practices that alleviated screwworm losses. However, a satisfactory solution was not developed until a new procedure for pest population suppression was investigated and developed. This involved the mass rearing of screwworms in the laboratory in an artificial medium consisting of ground meat, blood, and water. This development, plus detailed information on the biology, ecology, and dynamics of the insect, led to the concept of control by mass rearing and sterilization of insects for release to compete with the native wild flies. It was demonstrated in the laboratory that the males of the screwworm receiving about 4,000 roentgens of gamma rays emanating from radioactive cobalt were made sterile, yet, the males could compete successfully with normal males for mates. Normal females mated to such sterilized males will lay eggs, but the eggs do not hatch. Reared females treated with X rays or gamma rays are also sterilized and can be released along with the sterile males without interfering with the effect of the sterile male releases.

The laboratory studies were followed by field investigations which culminated in a field experiment to test the principle of screwworm population suppression by the release of sterile males on the island of Curacao. Sterilized flies were released on the 170 square mile isolated island at the rate of 800 per square mile per week beginning in August 1954. This rate of release exceeded the natural population enough to start a downward trend in the wild population. By maintaining a constant rate of sterile fly releases, the ratio of sterile to fertile flies progressively increased as the natural population declined. Within about 3 months or during a period of about four generations of the pest, no screwworm cases in animals were recorded on the island. The island has remained free of screwworms to date. The research on Curacao confirmed theoretical calculations of the expected effect of the sterile fly releases. It demonstrated conclusively the soundness of the principle of pest population suppression by the massive and sustained release of sexually competitive sterile organisms.

After the success on Curacao, a cooperative program for screwworm eradication in the Southeast was undertaken. A screwworm rearing factory was constructed and screwworms were reared, sterilized, and released at the rate of about 50–60 million per week. The program started in January 1958. By June 1959 the last screwworm case in livestock was recorded and the program was terminated in November of that year. The Southeast has remained free of screwworms since the population was eradicated. In carrying out the program, about 5 billion screwworm flies were reared, sterilized, and released in an area exceeding 50,000 square miles during a period of about 20 months. About 20 aircraft were used to distribute the flies each week. The total cost of the program was less than $10 million. According to estimates by the livestock growers, savings to the livestock industry since 1958 have exceeded $200 million.

Following the success in the Southeast a cooperative program was undertaken in 1961 for eradication of the screwworm in the Southwest. Permanent elimination of screwworms in the Southwest is not technically feasible because of continuous infestations in Mexico and Central America. The pest is highly mobile and is capable of flying several hundred miles in its lifetime. This was a substantially larger program and required the construction of a rearing plant having a fly production capacity of more than 150 million insects per week. These insects were sterilized and released in the southwestern infested area, beginning during the winter months when the pest was restricted in distribution and abundance. The program, which is still under way, is being carried out in cooperation with the Government of Mexico. It has been highly successful in spite of the constant threat of immigration of flies into the suppression area from untreated areas in Mexico. The suppression of screwworms attacking livestock, deer, and other wild animals is well over 99%. The cost of the program amounts to approximately $6 million annually. However, the livestock interests estimate savings of about $100 million per year.

Negotiations are under way by government officials of the United States and Mexico to develop a cooperative program of screwworm elimination in the United States and Mexico, and the eventual establishment of a barrier to reentry of the insect in Southern Mexico.

The sterility method that has proved so successful in dealing with the screwworm problem is under investigation as a possible means of eradicating or suppressing populations of other insects, including such major pests as tropical fruit flies, boll weevil, codling moth, pink bollworm, tsetse flies, and certain disease-carrying mosquitoes.

Another approach to successful management of populations that compete with man is evident in techniques for combating devastating food crop diseases. In 1946, a plant disease, called late blight, appeared suddenly and without warning on tomatoes throughout the Eastern United States. In one season, the total estimated loss was over 40 million dollars. This disaster led to establishment of regional research programs that have developed techniques to predict plant disease outbreaks and provide a basis for preventive measures for crop protection.

Through research on the biology of the late blight fungus and the effects of climate on development of the disease, scientists are able to forecast the occurrence of outbreaks of disease. Rather than apply routine and often unnecessary protective sprays or dusts, growers are notified when conditions are favorable for disease and apply protective measures as needed.

Forecasting has been extended to several plant diseases that occur over wide areas when pathogen populations build up to epidemic proportions. Disease warning systems for potato and tomato late blight, sugarbeet leafspot, apple scab, several vegetable downy mildews, and' various cereal diseases are now in operation in various states and regions of the United States. Based on research on prevailing weather patterns and plant disease spread, cooperative efforts with Mexico and Canada have been under way for several years. Forecasting buildup of plant pathogen populations is estimated by the USDA to cost approximately $100,000 annually. This one cooperative research effort has resulted in a savings to United States growers of at least 1 million dollars per year; a savings that is reflected in the low cost of food items for the consumer. Forecasting methods developed in the United States have been adopted in India, Israel, Japan, and Germany and are being initiated in the CENTO countries of Turkey, Iran, and Pakistan.

Similar to agents of human disease, plant pathogens exist in genetically identifiable races or types. One of the most destructive is the cereal rust fungus that attacks wheat, oats, barley, and rye causing drastic reductions in yield per acre, total grain production, and income. In 1950, the cereal rust fungus destroyed 25% of the wheat crop of the Northern United States. This loss was due primarily to a relatively new rust race (15-B) to which all commercial varieties of wheat were completely susceptible.

The effects of this overwhelming population of one fungus could hardly be considered a success story. On the other hand, without the research contributions made by plant scientists who constantly wage a war against diseases and pests, it is doubtful that the United States would have produced sufficient small grain to meet even its own food requirements during those years.

There are more than 300 races or strains of wheat stem rust, some of which increase in quantity and eventually destroy even the most rust-resistant wheats. This forces plant scientists to carry on a continuing effort to develop varieties resistant to rust. Methods for controlling such a widespread disease once it attacks have proved impractical. Rust control by breeding requires both national and international efforts. Wheats from the United States are exposed each year to dangerous rust races in special plantings grown throughout the world. In 1969 the International Spring Wheat Rust Nursery was grown at 59 locations in 30 countries and on 5 continents. The annual distribution of rust race populations is monitored throughout the United States, Mexico, and Canada to detect dangerous new races which can attack present commercial varieties, and then to begin development of new wheats to meet the new threat.

Plant scientists go to many parts of the world to secure new germ plasm to be used for breeding high yielding and resistant varieties. To illustrate the role of a single plant introduction in wheat breeding, in 1948 a wheat variety was collected from Turkey which on testing was found to possess certain qualities considered undesirable for production in this country. Consequently, it was laid aside and received little attention until the early 1960's. During this period, 1948–1960, stripe rust, another fungus disease, had become epidemic and was devastating to wheat production in the Pacific Northwest. Further tests with the Turkish variety determined that it had good resistance to several races of this rust and to bunt, still another serious wheat disease. Thus, after being ignored or neglected for a number of years, this introduction has become one of the prominent breeding lines used in Montana, Idaho, Oregon, and Washington.

The cereal disease problem affecting the Great Plains and Northwest small grain crops indicates the efforts required by plant scientists to maintain and increase crop plant productivity. The end result is better yields, a more dependable supply of grain, and increased farm and business activity. A comparison of Upper Midwest small grain yields in the 1950's when rust caused heavy losses with the 1960's when newer rust resistant wheats prevented widespread losses is striking. The results of this one research effort alone have been credited with

contributing more than 1 billion dollars to the Upper Midwest economy from 1962 to 1965.

Successful regulation of pest populations at satisfactorily low levels on a permanent basis by the use of natural enemies has been achieved now in an impressive number of cases. Such applied biological control has been emphasized most with agricultural insect pests and weeds but also it has been directed successfully against other groups of organisms such as rabbits in Australia. The method in its classical form involves the introduction of a new density-dependent mortality factor (i.e., an exotic natural enemy) into a pest population in order to achieve a new population balance at reduced densities.

The first great success literally saved the young citrus industry in California in 1890. The cottony-cushion scale, an invader from abroad, was causing such widespread damage that citrus orchards actually were being pulled out. Real estate values declined drastically. The discovery of the prey-specific predatory vedalia beetle in Australia and its colonization in southern California in 1888–1889 resulted in complete control of the scale within about 1 year. Because it involves natural balance between predator and prey, such control is permanent and benefits to man accrue year after year. Otherwise costly chemical treatments would be necessary indefinitely. The substitution of biological control of pests for chemical control has an additional great advantage. It reduces environmental pollution from pesticides. This cannot be expressed in dollars saved but the contribution to human health and values is obvious.

Since 1890, importation of exotic natural enemies to achieve biological control has been directed against some 223 species of insect pests, worldwide. Although research support has been limited and some of these attempts were merely incidental projects, an appreciable degree of success has been attained against 120 pest species. About three-quarters of all successful cases have resulted in either substantial or complete control of a major pest, so that need for chemical control was greatly reduced or was permanently eliminated. Worldwide, this includes complete and permanent control of 42 species, substantial control of 49 species, and partial control of 29 species. Additionally, several pest species have been controlled in from 5 to 30 countries, so that altogether there are about 103 cases of complete biological control, 95 cases of substantial control, and 55 of partial control. The importance of partial biological control should not be underestimated; it may mean a reduction of 50% in the need for chemical control.

There also have been some spectacular successes in biological control of weeds such as with the prickly pear cactuses in Australia and the Klamath weed in California. In the former case, several million acres, in the latter case, tens of thousands of acres of infected rangeland that had been rendered useless for grazing were restored to useful production as a result of complete biological control being achieved. Other methods of control had failed or were impractical. All told today 27 species of weeds have been controlled to greater or lesser degrees in 12 countries.

California has been a leader in biological control research for many years and some remarkable successes have been achieved there. It is conservatively estimated that in California alone well over $200,000,000 has been saved the agricultural industry of the state in the last 47 years (i.e., since the University of California became responsible for biological control research work in the state). This does not include the early success against the cottony-cushion scale that alone would involve savings of many millions of dollars. World estimates are virtually impossible to obtain. Another index is furnished by the Commonwealth Bureau of Biological Control, which spent $2,400,000 in various countries in 40 years up to 1966, with resultant net savings to agriculture of some $11,500,000, and this did not include savings from several successful cases for which gains were not financially documented.

Even though man has been associated with rats for thousands of years, he has never devised truly adequate means of controlling these pests. Progress in the concept of environmental control and recently discovered rodenticides has given man important but not all-powerful weapons, by which he can reduce rat populations.

In the 1940's a series of studies was made on the rat population in Baltimore. These studies demonstrated several important points: *1)* the growth of a rat population follows the basic principles which had been discovered in the 1920's and 1930's for other species of animals. *2)* Populations, in general, increase to a level determined by the environment, that is, they increase to the limit set by the "carrying capacity" of an environment. *3)* The chief environmental factors determining the urban "carrying capacity" of rats are the availability of food and shelter.

These studies demonstrated that rat populations decimated by trapping or poisoning usually reestablished themselves in a relatively short time. They further showed that by altering the carrying capacity of neighborhoods, by reducing the availability of food and shelter through proper disposal of garbage and trash, the rat populations decreased without the use of traps or rodenticides.

The interplay between social attitudes, economics, and governmental organizations makes the initiation and maintenance of widespread and intensive campaigns of environmental control very difficult. Most workers in the field today agree that anti-rat campaigns should involve both environmental control and supplementary use of rodenticides.

Like the proverbial better rat trap, the perfect rat poison has been difficult to attain. In the early 1930's some investigators were concerned about sick cows. But not until a farmer arrived at a laboratory with a dead heifer, a milk can full of blood that would not clot, and 100 pounds of spoiled sweet clover that had been fed to his cattle was a clue found to the strange hemorrhagic disease in cattle.

After 6 years of intensive work, Dicumarol was isolated. This was the first of many compounds to be discovered that slowed the clotting rate of blood. These chemicals, because of their potency, were of immediate medical importance in treating patients with circulatory problems.

By 1950 a Dicumarol derivative had been adapted for rodent control and was named warfarin after the Wisconsin Alumni Research Foundation, which had supported the basic research. Warfarin revolutionized rat control, just as penicillin and other antibiotics had revolutionized medicine. When consumed by rodents in small amounts over several days, delayed clotting time, breakdown of blood vessel walls, and internal bleeding resulted in their death. Because of its cumulative action, it could be used readily in homes and factories and on farms; and even accidental consumption was not hazardous to children and most farm animals and pets.

Dicumarol found ready use in medicine, but some patients failed to respond under its therapy. Eventually, it was determined that some individuals inherit a "resistance" to this chemical. About the same time, government officials in Europe were learning that rats, too, could be resistant to warfarin. Rats and mice which cannot be killed with the usual doses of this rodenticide have been found in Denmark, Germany, Holland, England, Wales, and Scotland. Other kinds of poisons must be used in areas where these resistant rats occur, and so the search for the ideal rodenticide continues.

1-(1-Naphthyl)-2-thiourea (ANTU) is another classic rodenticide, but its history is devious. Most high school biology students have studied the inheritance of taste with a widely available compound, phenylthiourea. Some perceive a bitter taste; others have no reaction. In 1939 an investigator wondered whether rats would show this same differential response. His test animals exhibited no response when a minute amount was placed on their tongues, but the next morning all were dead. Then during World War II sources of red squill, widely used as a rat poison, were cut off, and an effective substitute was badly needed. In 1942, on the basis of the test information, over 200 related compounds were tested before one was selected for its ease of manufacture and high toxicity to Norway rats. It was given the chemical acronym ANTU, was finally patented in 1945, and was especially important in the decade before the anti-coagulants swept onto the scene.

Pharmaceutical companies create and analyze large numbers of candidate chemicals in their search for new drugs. In the screening process, many animals are tested. In one such routine test with anti-inflammatory compounds, the investigators noticed that rats and only rats, died quickly after eating small quantities of a new chemical. Subsequent tests showed that this material was specific for rats and that it would not kill other animals no matter how much they ate. Here was a fast-acting, highly specific rodenticide—the closest approach to the ideal rat poison yet. However, under field trials, the rats detected the very bitter taste of this chemical, which now was called Raticate, and would not consume lethal amounts. Currently efforts are underway to use a process called microencapsulation (originally developed for carbonless office paper) to hide the taste and restore this rodenticide to the arsenal of tools available to fight rodents.

Another chemical, termed Silitrane, has been brought to the field trial stage. While it is not rat specific, it breaks down quickly after being eaten; thus there is no hazard to a cat or other animal that could eat a dead rat. Here, too, is another rodenticide with some highly desirable characteristics that evolved quite incidental from other developments.

The introduction of warfarin (and subsequently related anticoagulants) 2 decades ago was understood by many to signal the end of rodent pests. Intensive research and development of new rodenticides as well as training of scientific and technical personnel in rodent control slowed down or ceased. Federal funds in support of such programs dried up.

Man had to learn the hard way that warfarin, like penicillin and DDT, was not a simple answer to a complex problem. Additional rodenticides and techniques were needed. But even more, it was necessary to learn how to change the environment so as to eliminate the food and harborage on which the rodent depended. By doing this, rat and mouse populations could be permanently reduced and rodenticides then could be used as valuable supplements. This philosophy was vividly demonstrated in Baltimore in the 1940's, but even today people have trouble really accepting it.

SELECTED ADDITION READING

1. BEITER, C. B., M. SCHWARTZ AND G. CRABTREE. *5-p-Chlorophenyl silitrane, a new single-dose rodenticide.* Rahway, N.J.: M & T Chemicals Inc., 1970, 8 p.
2. COOLEY, D. G. Men behind the medical miracles. *Today's Health Magazine,* January, 1959.
3. DAVIS, D. E. Urban rodent populations. *Seminar on Rodents and Rodent Ectoparasites* 1966, 61–64.
4. DE BACH, P. *Biological Control of Insect Pests and Weeds.* New York: Van Nostrand-Reinhold, 1964.
5. EMLEN, J. T., A. W. STOKES AND C. P. WINSOR. The rate of recovery of decimated populations of brown rats in nature. *Ecology* 29: 133–145, 1948.
6. FENNER, F., AND F. N. RATCLIFFE. *Myxomatosis.* Cambridge: Cambridge Univ. Press, 1965.
7. GOULD, R. F., ET AL. New approaches to pest control and eradication. *Advan. Chem. Ser. 41.* 1963, 74 p.
8. HANSBERRY, R. Prospects for nonchemical insect control—an industrial view. *Bull. Entomol. Soc. Am.* 14: 229–235, 1968.
9. JACKSON, W. B. Anticoagulant resistance in rodents. *Pest Control* 37(3): 51–55; (4): 40–43, 1969.
10. KNIPLING, E. F. Sterile-male method of population control. *Science* 130: 902–904, 1950.
11. KRIEGER, C. H. Compound 42—a new anticoagulant as a rodenticide. *Pests and Their Control* 17: 24, 26, 28, 1949.
12. LEE, J. O., JR. Outlook for rodenticides and avicides registration. *Proc. 4th Vertebrate Pest Conf.,* 1970, p. 5–8.
13. LINK, K. P. The discovery of Dicumarol and its sequels. *Circulation* 19: 97, 1959.
14. O'REILLY, R. A., AND P. M. AGGELER. Studies on coumarin anticoagulant drugs: initiation of warfarin therapy without a loading dose. *Circulation* 38: 169–177, 1968.
15. O'REILLY, R. A., P. M. AGGELER, M. S. HOAG, L. S. LEONG AND M. L. KROPATKIN. Hereditary transmission of exceptional resistance to coumarin anticoagulant drugs. *New Engl. J. Med.* 271: 809–815, 1964.
16. OVERMAN, R. S., M. A. STAHMANN, C. F. HUEBNER, W. R. SULLIVAN, L. SPERO, D. G. DOHERTY, M. IKAWA, L. H. GRAFT, S. ROSEMAN AND K. P. LINK. Studies on the hemorrhagic sweet clover disease. XIII. Anticoagulant activity and structure in the 4-hydroxycoumarin group. *J. Biol. Chem.* 153: 5–24, 1944.

17. *Pesticide Handbook—Entomology*, (22nd ed.), edited by D. E. H. Frear. State College, Pa.: College Sci., 1970, 283 p.
18. RICHTER, C. P. Experiences of a reluctant rat-catcher. The common Norway rat—friend or enemy? *Proc. Am. Phil. Soc.* 112: 403–415, 1968.
19. ROSZKOWSKI, A. P. Comparative toxicity of rodenticides. *Federation Proc.* 26: 1082–1088, 1967.
20. ROSZKOWSKI, A. P. The pharmacological properties of norbormide, a selective rat toxicant. *J. Pharmacol. Exptl. Therap.* 149: 288–299, 1965.
21. SLADEN, B. K., AND F. B. BANG. *Biology of Populations.* New York: American Elsevier, 1969, 449 p.
22. SIMMONDS, F. J. The economies of biological control. *J. Roy. Soc. Arts* 1967.
23. WHO Expert Committee on Malaria, 13th Report. *World Health Organ. Tech. Rep. Ser.* 357 1967, 59 p.

Chapter VII

Environmental hazards

JOHN J. HANLON, M.D.

Human health has always been dependent on the relationship of man to his environment. The human being, like other animals, is constantly being transformed by natural forces in the environment that act on him, and he is continuously transforming his environment. Human beings, other animals of all sizes and shapes from the sizable whale to the microscopic amoeba, bacteria that are smaller than the smallest animal, viruses tinier even than bacteria, and all plants of different sizes and shapes —all these living organisms are part of a dynamic system in which they continuously interchange matter and energy with the world about them (their environments). The human being, for example, takes into his body solid and liquid matter from his environment and puts solid, liquid, and gaseous waste into his environment; he exchanges gases with his environment and he even contrives machines and gadgets that also exchange matter and energy with their environments. Other living organisms may get into man's body. Some act for his physiological benefit while others cause trouble. In addition, man both benefits from and becomes afflicted by the nonliving forces of his world. *All* of these interactions with the environment, including the great changes in the past 30 years, are involved in the relationship of man's environment to man.

Man's relationship to his environment necessarily must be regarded in two contexts: *1)* the elements of the natural environment that are hazardous to his health and

243

safety, and *2*) man's actions within his environment that themselves threaten his health. Within the first context are biological agents of disease or injury, including micro-organisms, noxious plants, and toxic or physically harmful animals; weather, including violent storms, cold, and thermal stress; natural radiation from the soil or from the atmosphere; geological perturbations such as earth-quakes, volcanic eruptions, and floods; and certain naturally occurring chemical substances. The second context consists of hazards to man's health which result from his own actions and maladaptations of the natural environment. This second context of human hazards brings the biological scientists into a confrontation with conditions even more complex than those that comprise health threats of "natural" origins.

Man-made hazards include such things as accidental injury, suicide, and homicide; injury and genetic damage from ionizing and nonionizing radiation; poisoning from industrial, agricultural, and therapeutic chemicals; noise-induced hearing loss; and health threats from polluted air, land, and water. These health problems are direct though unintended results of social and technological advancement or aberrant human behavior or both.

The automobile is an outstanding example of a social and technological advancement that has brought with it a staggering toll in accidental death and injury, pollution of air, water, and land, and unmeasured hearing loss from noise as well as some of the most hazardous occupational environments such as oil drilling and highway construction. Elimination of the automobile as a threat to health and life cannot be achieved either by its removal from society or by "immunizing" mankind against its effects. Inevitably, a trade-off must be reached where our chosen systems of transportation provide sufficient benefits to justify whatever toll results in terms of health effects. Similar acceptable cost–benefit decisions will have to be made with respect to radiation, pesticides, drugs, industrial chemicals, and most other hazards in this category.

Aberrant human behavior underlies many health hazards. Obvious examples are the steadily rising incidence of death from suicide and homicide. More subtle but significant is the inexplicable risk-taking associated with misuse of narcotics and alcohol, exposure to venereal diseases, and many sources of accidental injuries. An ever-increasing number of people in our society seem willing—even eager—to risk the almost certain health threat of addiction or accidental death or injury so as to achieve whatever the benefit that risk offers them. There is much fancy and little fact to aid the life scientists in reducing health threats related to such behavior.

The unique nature of human and environmental hazards, especially those generated by man, is such that many complex social, economic, and political factors bear heavily on them, their genesis, their study, and their solution. Although the net result of such hazards is an unwanted effect on man's health and well-being, the traditional training and experience of the life scientist is insufficient to cope with these problems single-handedly. By introducing the element of human action—either purposeful or by indirection—the health scientist finds it necessary to form an alliance with engineers, physical scientists, behavioral scientists, economists, educators, and political scientists if prevention or control of these human hazards is to be achieved.

The successful development of the antipoliomyelitis vaccine against a biological hazard was an excellent example. Health scientists recognized the health hazard and the need to eliminate it. They conceived of and successfully pursued the biological means of accomplishing this. But the actual fulfillment necessitated working with physical scientists to develop the production equipment; with engineers and industrial designers to develop a high-speed hypospray for safe, sterile, rapid, mass application of the vaccine; with behavioral scientists and educators to motivate individuals and groups and to understand and overcome the objections and misunderstandings of some; with economists to obtain the significant amounts of funds required for polio research and application of its results; with lawyers to provide legal protection; and with various elected and employed practical political scientists and public administrators to secure the necessary social and community sanctions. Throughout, from conception to successful application, the biological or health scientist was fundamental, but assistance was needed from the others. The total cost was great, but the benefit in lives and dollars was tremendous.

If one accepts the conditions of trade-off improvements as opposed to absolute prevention and need for multidisciplinary involvement, it is necessary to reexamine the criteria by which program priorities are established. Eight such criteria may be listed: 1) mortality rates, 2) morbidity rates, 3) financial impact on national economy, 4) utilization of hospital space, 5) numbers and availability of health personnel, 6) types of control measures available for attacking the problem, or the ease with which they can be attained, 7) available or obtainable human and financial resources, and 8) public receptivity to the idea and method of controlling the hazard.

None of the hazards to health considered here meets all of these criteria. Thus accidents and occupational injury and illness should receive high priorities in terms of the first four and possibly the fifth criteria but should probably be rated low on the last three.

With respect to scientific breakthroughs, the widest possible variance exists among the hazards considered. They range from vast strides and control over many communicable diseases to meager beginnings on the subject of noise. In a very real sense accomplishments serve as a measurement of both society's concern for the problem and its willingness to invest in its solution. Even more specifically, progress in controlling human and environmental hazards reflects the extent to which political leaders are motivated by physicians, scientists, and the general public who may believe that a particular hazard can and must be controlled and eliminated.

History describes the sequence: the first step is taken by a scientist who believes that a hazard might be understood and controlled. Then, after successful research has shown what can be done, concerned citizens and their political leaders are prompted to apply the results.

The past 3 decades have witnessed tremendous advances in medicine, science, engineering, and technology. In the United States varying degrees of affluence and high standards of living have been achieved for a large proportion of the population. The chief health problems of the past—yellow fever, plague, malaria, cholera, typhoid, dysenteries, and many others—have all succumbed to the application of research. Unfortunately, we now find that many of the recent technological advances have given rise to new and unfamiliar limitations to human life and happiness. Population increase, road construction, and transportation improvement have led to grotesquely increased crowding and its accompanying stresses.

An ever increasing barrage of new chemicals reaches man's body through food and water consumed, air breathed, and drugs taken therapeutically or otherwise. Streams and lakes have become contaminated with biological, agricultural, industrial, and mining wastes. Rapidly increasing amounts of garbage, solid wastes, and discarded vehicles multiply the opportunities for insect and rodent breeding. A crescendoing cacophony is affecting the auditory perception of an increasing number of persons. Ionizing and nonionizing radiation are a growing part of man's environment. It is used not only in nuclear testing but also in industry, laser and microwave technology, and in electronic products in the home as well as in devices for medical diagnosis and treatment. Each of the various environmental hazards is becoming increasingly significant in its own right, but the implications in terms of total "body burden" is yet to be comprehended.

Thus, the modern environment is dangerous on two accounts: it contains elements that are clearly noxious and stressful, and it has been changing so rapidly that man appears unable to make proper adaptive responses to it. The rate of technological change is accelerating every year; most changes demand a human adaptation. A complex of physical, mental, and emotional stresses of human life are resulting from increasing urbanization. Small but long-term exposures to many environmental stresses, hazards, and contaminants cause many previously little understood chronic ailments, disabilities, and even death. There is only now a beginning realization of the complexity of the environment and of the various factors within it that may act together to reinforce each other (synergism) or to oppose each other (antagonism). Those who are today concerned about human health and welfare must learn more and more to assess the problems of their own special areas of interest from an overall, a total, a holistic viewpoint. It will no longer do for the problems of one area to be considered in isolation or separately from the problems of related areas.

Very little has been learned about the interactions within the human body and human mind of the various environmental stresses, although it is known, for example, that the intended effects of certain medications may be initiated, intensified, or otherwise transformed by certain other factors in the environment. Similarly, it is known that radiation, cigarette smoking, certain air pollutants, and various chemicals encountered in the occupational setting can have a multiplying effect in the development of pulmonary cancer. The little that is known about multiple hazardous impacts merely highlights the vast amount still to be learned about human response to the environment. Most of the vast and ever-increasing array of physical, chemical, biological, and psychological stresses to which the modern urbanized American is now being subjected did not even exist in his grandfather's time.

In 19th century America the sociological and hygienic leaders were primarily concerned with malnutrition, overwork, filth, and microbial contamination. Today's characteristic diseases result from economic affluence, chemical pollution, and high population densities. The increase in chronic and degenerative diseases is due in part to these environmental and behavioral changes and not, as often mistakenly supposed, due to an increase in human life expectancy. Most people do not realize that life expectancy after the age of 45 years has not increased significantly. A person 45 years or older is little better off than he would have been in 1900 or before, even if he can afford good medical care. It can be assumed that medical science will continue to develop useful techniques for ameliorating cancers, vascular disease, and certain other degenerative disorders. Undoubtedly, it will refine methods for organ transplantation and for artificial prostheses. But most of the conditions to be treated need not have occurred in the first place. Greater knowledge of environmental determinants, provided by more biomedical research, is the path to prevention and control—a better and less expensive way.

The protection of the public from hazards involves skills and knowledge far more extensive and varied than seemed sufficient even in the recent past. To accomplish this requires sound scientific understanding resting on intensive and extensive biological and biochemical research. Fortunately there is a growing public awareness of the need for improved environmental quality. Increasingly, as environmental deterioration has affected more and more citizens, their uneasiness has produced demands for safer and more sensible use of the environment, and an increasing willingness to provide whatever support is necessary for the restoration and preservation of the environment. This implies involvement which, in terms of universities and similar concentrations of expertise, brings one back to the subject of research. In the absence of increased attention to the study of the effects of environmental hazards on man, increases in disease rates and greater expenditures for avoidable therapy are inevitable.

The population of the nation is projected to increase to 250 million people by 1980 and over 300 million by 2000 A.D. Metropolitan areas probably will absorb nearly all the increase, and environmental problems will be intensified geometrically. Moreover, new and presently unsuspected environmental hazards will be identified. Without corresponding growth in environmental health protective activity, the future can only bring mankind increased physiological and psychological suffering.

An overview of where we stand and how we must proceed is pertinent. Guidance is provided by the successes in the combined areas of environmental biology and aerospace physiology. Man's space travel provided the impetus to bring together the knowledge of several disciplines in a new more positive approach to human health. Thus research on space vehicles has provided facts that are useful in occupational health programs and radiation protection procedures. Already these breakthroughs have begun to result in improved health through the establishment of realistic criteria for technological assessment. They have increased substantially the awareness of the holistic approach to the solution of environmental problems at every segment of society encompassing both professional and general citizenry. Among the "spin-offs," the many specialists in environmental health have necessarily become broader professionally and are now more effectively bringing together the health scientist and the urban planner, and the lawmaker and the toxicologist, to mention only two of innumerable examples. Other spin-offs worthy of mention are in the area of therapeutic medicine. Better health care delivery systems have evolved and the need for comprehensive health programs has been more clearly defined.

Against this background, two major breakthroughs during recent decades are worth special note. The first was the appreciation that small amounts of chemicals may become toxic after periods of accumulation or latency. The kind of biological and toxicological research that was carried out on beryllium, asbestos, benzene, and mercury exemplify this. It demonstrated some of the fundamental enzyme systems involved and showed that the rate at which the body acted on a chemical affected its toxicity. As a result, the basic mechanisms of toxicity became better understood as these biological and chemical processes were probed in depth. Safety evaluation techniques were improved and methods for studying the potential of chemical carcinogenesis, teratogenesis, and mutagenesis, were and are still being developed as a result of these research efforts.

The second major and concurrent breakthrough was that the scientific community became aware that there was inadequate information on the effects of these widespread chemicals (air pollutants, metals, persistent pesticides, additives, pharmaceuticals, and the like) in the biosphere. As a result, the research community is now expressing its concern that the total ecological threat may in some instances be more important than any direct toxic effect on humans. In sum, the major breakthrough was to appreciate the latent, insidious, and chronic effects of small amounts of chemicals in the environment, often over prolonged time periods, and their potential threat to future generations through their effects on the biosphere. Even this breakthrough is inadequate unless consideration goes beyond acute, easily recognized, clinical responses to chemical toxins, to consideration of genetic potentials.

The number of factors in the environment that may affect man are presented in Table 1. Some of these factors in proper form or amount are helpful or even necessary to life, others are hazardous. Consideration of a number of them, with some consideration of significant breakthroughs and further needs, follow under three categories of hazards—biological, chemical, and physical.

BIOLOGICAL HAZARDS TO MAN

The biological hazards to human health are mostly those posed by bacteria, viruses, and other microorganisms and parasites. Since the advances against them have been discussed in other chapters, they will be considered here in a general way. Those microbiological accomplishments directly related to human health include modifications of man's susceptibility to disease or of his recuperative ability (better diagnostic and immunizing methods and better antimicrobial drugs); modification of his environment or of his relationship to his environment (more knowledge of the agents that cause or induce infectious disease and how to control them; better methods of insuring that food, milk, and water are free of disease-producing microorganisms; more aseptic homes, public institutions, and food distribution industries; better understanding of community health

TABLE 1. *Classification of environmental factors that may affect man*

Life Support
  Food
  Water
  Oxygen

Physical
  Mechanical
  Acoustic
  Electrical
  Magnetic
  Thermal
  Particulate
  Ionizing Radiation

Biological
  Microorganisms
  Toxins
  Biological Wastes
  Biological Antagonists
    Animal
    Plant
  Allergens

Psychosocial
  Crowding
  Demands
  Physical time
  Biological time
  Cultural time

Chemical
  Inorganic
    light metals and their compounds
    transitional and rare earth metals and their compounds
    heavy metals and their compounds
    nonmetallic elements and their compounds
  Organic
    acyclic hydrocarbons including alkyl compounds
    carbocyclic compounds
    halogenated acyclic hydrocarbons
    heterocyclic compounds
    organic phosphorus compounds
    organic sulfur compounds
  Product Complexes
    combustion products
    macromolecular products
    industrial wastes
    agricultural wastes (including fertilizers, pesticides and
      herbicides)

problems; and better control of disease in domestic and wild animals). Those accomplishments that have an indirect relation to human health include, for example, improvement in worker productivity with consequent increase in plant and animal food production. During the past 30 years microbiological research could have been even more effective had there been more complete communication between scientists, administrators, and agency representatives and less unnecessary duplication of research activities, whether funded by federal, local, or private sources.

Information exchange is vital to success and economy of applied research in the field of microbiology as related to health and disease. There has been little planned exchange of information other than through publication of scientific papers and presentations at scientific meetings. One notable exception is the Arbovirus Information Exchange, which is issued three or four times a year to scientists in 235 institutions. This, plus an annual meeting in conjunction with the Society of Tropical Medicine and Hygiene, minimizes unnecessary duplication and directs attention toward critical problems.

Microbiological research in the past might have been more effective if there had been a federally sponsored Expert Committee for each of the major diseases such as malaria and tuberculosis. The Armed Forces Epidemiological Board has had outstanding success with this kind of system, providing an example that should be followed by others, including the Public Health Service. Another deficient area of communication lies between the researcher and the tax-paying citizen, whose support for research could be retained if he had understandable information on research activities and accomplishments.

The following achievements may be anticipated during the next decade in this fight against microbiological hazards: 1) more effective and less expensive methods for disposing of human wastes thereby reducing water pollution from this source and reducing the biological usage of oxygen in rivers; 2) better methods of controlling disease caused by microbes in agricultural products (plants and livestock) leading to an increase in quality and quantity of food; *3*) more specific drugs for use against microbes, fungi, and certain viruses that cause human disease and disorders; *4*) more effective immunizing agents for human and animal use, including attenuated live virus vaccines administered orally or into the nose–throat region; *5*) determination of the causative agents of various types of cancer and their role in the pathological origin of cancer; *6*) improvements in the control of those diseases transmitted from animals to man, diseases such as Rocky Mountain spotted fever, plague, brucellosis, hemorrhagic fevers, encephalitis, rabies, and the like; *7*) more accurate and less expensive diagnostic microbiology due to automating and reducing the size of clinical diagnostic procedures; *8*) better understanding of immune mechanisms leading to significant improvement in the prevention and management of disease problems of the very young and the very old, particularly problems such as multiple sclerosis, rheumatoid arthritis, and disease states of the blood (blood dyscrasias).

While improvements in man's control of the biological hazards of his environment can be expected to continue, these advancements should be accelerated and made more effective by similar improvements in the kinds of communications needed for planning and coordination of research and development activites.

CHEMICAL HAZARDS TO MAN

The chemical hazards to which man is exposed exist in all three aspects of the environment—air, water, and soil. Although a particular chemical hazard initially may be limited to one of these, dispersion to one or both of the others may occur. This is especially apt to be the case for persistent pollutants. Agricultural chemicals present a classic example. As a result, a fourth source of environmental chemical hazard should be considered in the form of food, in the sense that it may reflect possible concentrations from air, water, or soil.

With reference to chemical hazards in terms of implications to human disease, perhaps the major advance of the last 3 decades has been the realization that spon- taneous diseases, for which no cause can be found, are in fact induced by environmental pollutants. Thus, certain cancers and deformities may be largely explicable in terms of exposure to certain specific environmental pollutants. More recently, there has been growing realization of the importance of chemical induction of genetic changes. Apart from preformed agents that may cause cancer (carcinogens), induce morphological abnormalities (teratogens), or bring about genetic changes (mutagens), there is a growing interest in the possibility of their synthesis in the body from simpler chemical compounds normally present in the environment.

Although the word pollutant is often restricted to synthetic industrial chemicals, there are four broad categories of other important chemical pollutants: the first group consists of natural chemicals in excess, and includes nitrates that are normal dietary components. At high concentrations in food or water, nitrates can cause a blood disorder (methemoglobinemia) in infants. Also, nitrites, as reduction products of nitrates, may interact with secondary amines to form nitrosamines. Some of them are carcinogens, teratogens, or mutagens.

Toxic chemicals in fungi on plants or in crop plant products comprise the second group, of which aflatoxins and cycasins are notable examples. The yields of these toxins can generally be considerably influenced by technological factors, such as conditions of harvest, storage, and processing. The third group consists of complex organic and inorganic mixtures, such as air and water pollutants, which comprise a wide range of undefined as well as defined components.

Finally, there is the group of synthetic chemicals— agricultural chemicals, notably pesticides and fertilizers; food additives; fuel additives; household chemicals; industrial chemicals; and in a somewhat specialized class therapeutic and prophylactic pharmaceuticals, as well as habituating and addicting drugs.

Pollutants may have a wide range of adverse biological effects, from health impairment to death. The possibility that chronic toxicity is also manifest in immunological impairment or psychological disorders has yet to be determined, although there is already suggestive evidence to incriminate carbon monoxide in these regards.

Some pollutants produce any one or more of these types of reactions, and may interact with one another outside the body as well as inside the body to produce otherwise unanticipated synergistic toxicity.

Naturally occurring chemicals. There are many naturally occurring chemicals that are hazardous to man. The toxicity of some, mercury and arsenic, for example, was known in ancient times. In recent decades, information from a variety of research investigations has indicated that many others play a role in human health and disease. Some trace elements are an important part in human nutrition and in many physiological reactions. But excesses and deficiencies of certain metals such as cadmium, selenium, chromium, lead, copper, zinc, lithium—or of radiation—may be related to major degenerative diseases such as heart disease, muscular dystrophy, diabetes, multiple sclerosis, cancer, mental illness, or congenital malformations. There is a growing awareness that many diseases have geographic patterns which seem to relate somehow to the physical environment and its different geologic metal-containing areas. This awareness has led to a corresponding recognition of the need for interdisciplinary teamwork among the geochemist and the medical scientists.

Some of the observations that have resulted from this multipronged investigative effort are described briefly herein.

For a long time *arsenic* was believed to produce cancer but recent research with animals and lack of evidence in man has dispelled this notion in the United States. Major sources of arsenic intake are food, especially crustacea, and beverages, particularly wines. Except for some isolated sites, the arsenic content of water supplies is extremely small. But in Europe there have been claims that arsenic does induce cancer. Arsenic has long been known to counteract selenium toxicity and it has been added to poultry and cattle feeds to suppress selenium toxicity in areas where selenium content of soil and water is high. But in hamsters selenium does not suppress the monster-forming effect of high levels of arsenic. It is possible that the different United States and European views of arsenic as a cancer-inducing agent may result from differences in exposures to these two antagonistic elements in the two regions of the world and to the quantities and chemical forms of these substances taken in.

Certain tumors (mesotheliomas) have been known for decades but it was not until 1965 that *asbestos* was found to be a causative agent, most likely because it is a source of certain trace metals and certain chemical compounds (such as benzpyrene). Benzpyrene is a ubiquitous air pollutant, a prominent constituent of tobacco smoke, and of charred and smoked foods; the particular trace metals are associated with certain types of asbestos fiber (chrysotile). There is some indication that the cancerous growths are brought about by an interaction between benzpyrene or benzpyrene compounds, an enzyme, and the particular trace metals. The chief source of non-occupational asbestos exposure is not yet actually known; possible sources are asbestos water pipes, asbestos filters for beer, asbestos brake linings, asbestos household iron-holders, asbestos used in some cigars, and asbestos in water supplies (possibly the most continuous source).

Beryllium causes a serious disease, berylliosis, a name only about 30 years old. Also it is known to induce tumors in animals but has not been found to do so in man, although vigilance is being maintained regarding this. Beryllium poses no threat to health from water or food sources, for it is poorly absorbed from the gastrointestinal tract and is highly insoluble in water either in or outside the body. The potential environmental threat is from the vicinity of beryllium production plants, for the use of beryllium as an industrial material is increasing rapidly.

Another chemical hazard is the trace element *cadmium*. In 1965 it was shown that the kidneys of persons dying from hypertensive complications had increased amounts of cadmium or increased ratios of cadmium to zinc, as compared with persons dying from other major diseases. This discovery has been substantiated experimentally in animals. The chief sources of cadmium are believed to be drinking water and food grown in soils containing cadmium from certain fertilizers; from beverages that have been in contact with galvanized zinc (container coatings) which contain small amounts of cadmium; and from vegetables, coffee and tea (all of which usually contain small amounts of cadmium). Drinking water is not thought to be a large source. Whether or not cadmium induces renal hypertension seems to depend also on the intake of zinc and selenium, antagonists to cadmium. In hamsters monster-formation induced by cadmium is antagonized by selenium. The highly lethal effect of cadmium on fish is increased (synergized) by cyanide.

A very important advance was made in 1960 when it was found, in some forty-eight states of this country, that water with relatively more dissolved salts, so called "hard water," was associated with lower death rates from cardiovascular and coronary heart disease. Not all states showed this correlation and the correlation was somewhat better for coronary heart disease in white men 45 to 64 years old, by state and in the 163 largest cities that had about 58% of the total national water supplies. Mortality from all other causes showed no such correlation. Prior to this discovery, state-to-state variations in cardiovascular deaths were unexplained on either a dietary, racial, or social basis. Evidently something in hard water, or the lack of something in soft water (relatively low in dissolved salts), influences the death rate. Of the various trace elements for which correlations have been worked out, only cadmium and *chromium* (Cr^{3+}) have been revealed as trace elements affecting cardiovascular disease. As pointed out in another chapter, this discovery raises the possibility that perhaps in time the prevalence of cardiovascular disease may be decreased by adding certain trace elements to water supplies.

Lead poisoning has been known from ancient times, but its clinical manifestations were classified about 110

years ago. Lead exists in many forms in man's environment and its industrial use is practically ubiquitous.
Occupational safeguards have limited its harmful effects,
but a special hazard is the use of tetraethyl lead as an
antiknock and power-increasing agent in gasoline. Tetraethyl lead is highly toxic when inhaled or absorbed
through the skin, and predominantly causes cerebral
or central nervous system symptoms. It pollutes the
atmosphere. The toxic lead-bearing household paints
have been largely eliminated but still, in this country,
thousands of small children eat dangerous amounts of
flaked household lead paint, mostly in the older and
poorer parts of large cities; some die and others develop
mental retardation, cerebral palsy, convulsive seizures,
blindness, and various other disorders. Research is being
directed toward safe and effective physiological delead-
ing procedures, and recently chromium as a trace element in food appears to decrease lead toxicity in some
animals—and presumably in man. Evidence suggests
that chromium (trivalent) acts as an antidiabetic and
antiatherosclerotic agent through its role in sugar and
fat (glucose and lipid) metabolism.

Inorganic *nitrates* and *nitrites* are a health concern
because they contaminate drinking water. They are a
special hazard to infants up to 6 weeks of age; and also
to Alaskan Eskimos and Indians, those who have a particular genetic blood disorder. The young infants are
susceptible because they have not yet developed certain
metabolic enzymes, they have a more readily reactive
hemoglobin which decreases after birth, and they have
a small blood volume relative to fluid intake. Milk from
cows that drink nitrate-polluted water can be an added
threat to children. Although nitrates are used in preservation of some meats and fish, such food is not a common
part of infant diet. The toxic effect of nitrates in infants
is a blood disease and, if intake is not stopped, death.
Although in this country community water supplies are
monitored in accordance with standards set by the Public
Health Service, well water on farms has no such surveillance. Any increase in the use of nitrate fertilizers
and superphosphate wastes, both being sources of nitrites

produced by bacterial action, will necessitate even greater surveillance and increase the hazard of nitrates and nitrites in well water on farms.

Toxic chemicals in fungi and plants. Substances produced in the process of growth and development of certain forms of fungi, more commonly called molds, represent another group of toxic chemicals. Molds cause economic loss through deterioration of food fiber, induce plant diseases, and cause certain pulmonary and invasive diseases in man. Mold-damaged foods were once considered harmless and used as animal feeds. We know now that a few molds of grain and other foodstuffs produce toxic chemical substances (metabolites) called mycotoxins.

The oldest mycotoxin is known as ergot, from a fungus that infects cereal grasses, especially rye. Ergot has been the cause of serious epidemics in Western Europe and parts of Russia since the middle ages. The most recent outbreak of ergotism occurred in 1951 in a region in France. There many people were stricken, some died; a large proportion showed central nervous system symptoms such as hallucinations, depression, and self-destruction manias.

Among the substances (alkaloids) identified in ergotized grain are some related to LSD (lysergic acid diethylamide). One of these substances, ergotine, causes contractions of blood vessels (arterioles) and smooth muscle fibers, and is used medically to stop hemorrhage, especially after childbirth, to stimulate labor, in spinal and cerebral congestion, in paralysis of the bladder, and in diabetes (diabetes mellitus). Ergotamine tartrate inhibits certain nerve endings (sympathetic) and has been used to treat migraine.

Other mycotoxins include the aflatoxins, produced by molds on peanuts and other agricultural products. Even in low concentrations aflatoxins cause acute intoxication, liver damage, and cancer of the liver (hepatoma) in animals. Certain molds on wheat allowed to remain in the fields during the winter produce poisons that cause a blood disorder (alimentary toxic aleukia). In 1944, in a district of the U.S.S.R., this disorder affected about a tenth of the population. Another mold on millet can cause an epidemic of excessive urinary secretion (epidemic polyuria). The paucity of knowledge regarding these types of hazard points up the great need for future research efforts.

Industrial wastes. In addition to naturally occurring chemicals and plant and fungal poisons, complex organic and inorganic chemical mixtures made up a third group of environmental hazards. They arise as a result of the pollution of air and water by combinations of community and industrial wastes. Man is now beginning to learn that the atmosphere and waters of this earth, large though they be, are not limitless and cannot with safety be used as sewers and disposal dumps. Air, water, and consequently food cannot continue to be mistreated in this way without rapidly increasing danger to a rapidly increasing population in a limited world. Most people do not realize that the average person in this country takes into his body each day about 30 pounds of air, 4 pounds of water, and 3.5 pounds of food. The unused remains of these are returned to the environment as contaminants or as potential contaminants.

There has been a belated recognition that air pollution is one of the most important sources of chemical hazards, especially in highly urbanized and industrialized societies. More than 200 million tons of toxic material are released into the air above the United States each year, about 1 ton per person. About 60% comes from approximately 100 million internal combustion engines, the other 40% from sources such as factories, power plants, municiple dumps, and private incinerators. The pollutants include carbon monoxide, sulfur oxides, nitrogen oxides, hydrocarbons, and particulates. They produce ill effects in people either by short-term, high-level or by long-term, low-level exposures (see Table 2).

Short, high-level exposures cause acute reactions with increased mortality especially among the elderly and the chronically ill. There may be increase of respiratory infections, irritation of ear, nose, and throat, and impairment of physiological functioning. The more serious effects that may lead to death include an increase in

TABLE 2. *Estimated atmospheric emissions, U. S. 1968, 10^6 tons/year*

Source	Carbon Mon- oxide	Partic- ulates	Oxides of Sulfur	Hydro- carbons	Oxides of Nitrogen
Transportation	63.8	1.2	0.8	16.6	8.1
Fuel combustion in sta- tionary sources	1.9	8.9	24.4	0.7	10.0
Industrial processes	9.7	7.5	7.3	4.6	0.2
Solid waste disposal	7.8	1.1	0.1	1.6	0.6
Miscellaneous	16.9	9.6	0.6	8.5	1.7
Total	100.1	28.3	33.2	32.0	20.6

the severity of chronic illnesses such as bronchitis, emphysema, asthma, and heart attacks (myocardial infarction). Such an episode occurred in 1930 in a valley in Belgium; 100 persons were stricken and 63 died. In Donora, Pennsylvania, in 1948, 6,000 of a population of 14,000 persons became ill and, during the episode, the death rate was 10 times the normal. During a 5-day fog and smog episode in London in 1952 emergency bed service requests increased 2.5 times and approximately 4,000 deaths were attributed to the air pollution. These are only some of the more dramatic epidemics caused by polluted air. In this country more than 6,000 communities are considered to be affected by varying degrees of air pollution. Inhabitants who show asthmalike responses, the most frequent response to polluted air, tend to have an increase in β- and γ-globulins in their blood, and biochemical studies have shown that asthma patients also have increased γ-globulins.

The insidious long-term effects of low-level exposures to polluted air perhaps are even more significant, for these exposures result in slowly developing difficulties hard to define or measure. Chronic diseases of several kinds, including cancerous growths and genetic mutations, may be initiated, general body defense mechanisms impaired, and physiological function interfered with.

The protective cilia of the respiratory tract, for example, may be slowed down or destroyed by atmospheric sulfur oxides, nitrogen oxides, and ozone and this injury may be associated with and followed by other respiratory disorders and diseases. In the United States chronic bronchitis and emphysema are among the most rapidly increasing causes of death. One in twenty asthma patients is severely affected by air pollution. Lung cancer has been correlated with air pollution as well as with certain other factors, such as smoking habits.

Deaths from pulmonary carcinoma have been correlated with air pollution, taking into consideration a number of pertinent socioeconomic variables, such as smoking habits, economic status, and degree of urbanity. For example among English nonsmokers, a 10-fold difference was found between the death rates for cancer of the lungs for rural and urban areas. Adult British immigrants to New Zealand and to South Africa have been found to suffer a higher incidence of lung cancer than individuals of the same ethnic stock who were born in those two locations. Among Norwegians living in Norway where air pollution is low, the lung cancer rate is also low. Among Americans living in the United States where air pollution is heavy, the rate is twice as high. For Norwegians who have migrated to the United States, the rate is half-way between.

Correlations have also been found between air pollution and carcinomas other than pulmonary, such as of the stomach. In addition, deaths from cirrhosis of the liver have been observed to be higher in heavily air-polluted areas, probably because livers already damaged by alcoholism decompensate when exposed to toxins of polluted air. In the laboratory, injection of animals with trace amounts of extracts of urban atmospheric pollutants from some cities, especially those using predominantly solid fuel, produced a high incidence of tumors of the liver, lymphatic system (lymphoma), and lung (multiple adenomas). Such amounts would be inhaled in about 3–4 months by a resident of the cities involved.

Quite apart from frankly evident air pollution and smog, specific gases may be harmful to health. A study in Chattanooga, Tennessee, has linked relatively low levels of nitrogen oxides to the susceptibility of children to the Asian influenza. Evidence from California indicates that a concentration of carbon monoxide in the air of as little as 10 ppm for about 8 hours may result in impaired mental performance because of reduction of ability of the blood to carry sufficient oxygen to the brain. Such levels are common in many cities around the world.

Elevated levels of carbon monoxide have also been associated with the increased probability of motor vehicle accidents and with the inability of individuals to survive myocardial infarctions. While admittedly several other factors are involved, it is notable that death rates from coronary heart disease are 37% higher for men and 46% higher for women in metropolitan areas with high atmospheric pollution levels than they are in nonmetropolitan areas. Cardiovascular death rates are more than 25% higher for male Chicagoans between 25 and 34 years of age than for their counterparts in rural areas. The difference is 100% for men between 35 and 54, and nearly 200% for men between 55 and 64. Since 1940, genetically determined coronary artery patterns have been recognized as playing a significant role in vulnerability to coronary atherogenesis and myocardial infarction. Of the three branching patterns identified, hearts with a left coronary artery pattern are considered most vulnerable to fatal coronary artery occlusion, and those with balanced patterns, least vulnerable. These and similar findings, combined with those that associate atherosclerosis with carbon monoxide, indicate the inadvisability of certain types of individuals residing in areas of high pollution.

In the earlier discussion of lead as an environmental chemical hazard, passing mention was made of the special problems presented by the addition of tetraethyl lead to gasoline for antiknock and extra power purposes. The consumption of leaded fuels has increased tremendously—from 100,000 pounds in 1940 to 450 million pounds in 1967, an increase of 4,500-fold in 27 years. If a uniform distribution throughout the country of its 90 million cars and trucks is assumed, the potential exposure to each of the country's individuals (200 million) approximates 2.25 lb./year; certainly a frightening thought, particularly when lead distribution is not uniform, but is concentrated in areas of heavy traffic.

Confronted with the facts, several investigators made worldwide comparisons of lead concentrations in the blood and urine of man. These comparisons showed: 1) variations in body concentrations did not exceed 3.5-fold anywhere in the world; 2) lead values in primitive societies often equaled those in the metropoli; 3) lead values had not changed in 3 decades; and 4) there is no evidence that at current lead levels adverse effects of the health of man occur; urinary D-aminolevulinic acid levels of normal populations are well below response levels.

Inferences from the above blood and urinary lead levels are somewhat misleading unless one notes that they are average values. Blood lead values were measurably higher in certain groups, traffic policemen and persons living adjacent to busy highways. In none of these individuals, however, did lead values approach levels considered hazardous.

Nevertheless, the measurable elevations in blood lead found by repression analysis are interpreted as evidence of a measurable contribution to the body lead burden from the atmosphere, heretofore not considered possible of attaining significance in comparison with the daily intake from food and beverage. The evidence is made plausible by the postulation of body absorption of up to 50% of atmospheric lead generated in submicron particle size from motor exhausts. Experiments with animals suggest a need for the larger cities to establish an air standard for lead. At the time of this writing, Philadelphia and New York City set limits on atmospheric lead concentrations. However, this limit (5 $\mu g/m^3$) may prove not to be too conservative, in light of increased toxicity of lead in certain deficiency states observed in advanced age groups and the demonstration in man of the interference of lead in various endocrine functions.

While many other examples of the relationship between general atmospheric pollution and ill-health might be presented, the seriousness of the situation is perhaps most succinctly illustrated by a recommendation of the California Department of Health. They propose that physicians should estimate the contribution that local air pollution conditions may make to the outcome of a patient's illness, and that in areas with high air pollution, some patients may benefit from a preoperative period in a clean-air room or chamber before receiving general anesthesia.

Water pollution. Water is one of the prime necessities for life. Without it, survival beyond a few days is impossible. In addition to oral consumption modern man has another use or misuse of water: as a vehicle for the removal of human wastes, as a great cesspool, for an ever-increasing complex mixture of industrial and related waste materials. As a result, there have been repeated, extensive water-borne epidemics of bacterial diseases, such as cholera and typhoid fever. In the late 19th and early 20th centuries technological means were developed to remove these hazards from water with notable declines in case and death rates. Yet even today cholera and typhoid fever epidemics occur with alarming frequency.

The hazardous aspect of water has been recognized of late as being greater also because of its contamination by viruses. Thus current levels of bacterial purity of water must be maintained and improved, practical methods of removing disease-producing viruses must be developed, effective means of surveillance and control of chemical pollution must be developed and put into practice.

Man is beginning to realize that he is living in an environment replete with agents that induce, promote, and accelerate cancerous growths and genetic mutations. Such substances contaminate man's drinking water as well as urban air and other parts of his environment. Water pollution with hazardous chemicals may be expected to increase as a result of the development of new industrial plastics, chemical sterilizing compounds for use against pests, missile fuels, and other substances.

Many of these substances are unstable; they may break down in water before they become a significant health hazard. But a threat may be posed, according to some epidemiological evidence, by a combination of potentially mutagenic chemical agents with radioactivity in water sources.

The National Community Water Supply Study completed in 1970 made a wide variety of recommendations for the delivery of an adequate supply of potable and safe water. Required is strengthened legislation, forthrightly enforced, to limit or prevent the discharge of untreated human and industrial wastes into the streams and lakes of the nation.

With specific reference to chemicals, inorganic and organic, the study clearly evidences the need: *1*) to simplify and lower the cost of removing excess chemicals known to be dangerous to the public health; *2*) to improve systems to control undesirable concentrations of iron, manganese, hydrogen sulfide, and color as well as organic chemicals causing unpleasant taste and odor; *3*) to develop surveillance techniques or conditioning procedures to eliminate the deterioration in water quality between the time that the water leaves the community water treatment plant and the time it reaches the consumer's tap.

A first step is better monitoring of water quality. Instrumentation for routine continuous monitoring with computerized interpretation must be developed. As of now, the monitoring of drinking water of the United States employs satisfactory but very inefficient methods. A tremendous investment will be necessary if the proper tools and the proper skilled personnel are to be provided so the presence of potentially toxic materials in the environment can be identified. Only then can we begin the epidemiological studies necessary to identify the significance of various toxic substances in water.

The outstanding breakthrough of recent years in instrumentation has been the development of membrane filters for the collection of microorganisms and large molecules. The membrane filter has permitted many research workers to adapt the filter to a wide range of problems, and the uses to which this type of membrane can be put are still being explored. A similar development might come from the application of other kinds of membranes, particularly the cellulose acetate membranes being used in desalination technology, which have possibilities for the concentration of microorganisms and other contaminants. There are numerous voids in existing technology that do not allow measurement of current procedures. Current drinking water standards do little more than mention viruses, neglect numerous inorganic chemicals, and identify only the index that is to cover the entire family of organic compounds. A breakthrough required in the next decade is the development of similar concentration and identification techniques for viruses, and particularly the ability to grow the infectious hepatitis viruses in vitro.

With the trend toward multiple use of water sources, plus the complex types of chemical contaminants, new methods of surveillance and treatment are required. Among the new contaminants that must be dealt with are fertilizers, herbicides, fungicides, irrigation residues from agriculture, detergents from homes and industry, radioactive wastes from power plants, industrial, and research installations, a spectrum of heavy metals, a wide variety of salts, and numerous other materials. Many of these are not readily biodegradable, are unaffected by conventional treatment methods, and build up in water supplies. This makes necessary continuous spot-testing techniques based on a variety of approaches. Some measure particular characteristics such as biological or biochemical oxygen demand and acid concentration. Kits for determination of more than 100 different physical, chemical, or biological characteristics or contaminants are now available. There are also more versatile instruments such as the direct-reading colorimeter with which 20 or more different tests can be made. The gas–liquid chromatograph has proved its value in identifying traces of organic materials in water. Continuous water pollution monitoring of effluents from industrial plants can be carried out by ion-selective electrodes consisting of an ion-selective membrane sealed onto

the end of an insulating glass or plastic tube containing
an internal reference electrode of silver chloride or cal-
omel. The solution to be measured is placed in the tube
and a voltmeter measures the electrical potential developed between it and an external reference electrode
when both are immersed in a solution.

The need for knowledge about the health effects of
water-borne contaminants will require thorough investigation. The concentration levels at which numerous
contaminants, such as mercury, molybdenum, or selenium, cause adverse health effects must be determined.
Similarly, we must soon determine the effect of the long-
term ingestion of low-level concentrations of toxic, organic materials in water. Some of these toxic materials
are carcinogenic, some mutagenic, some teratogenic,
and some have other toxicological effects. Some of these
effects are only identifiable after many years of exposure,
and because the effects are little different from other
types of exposure and deterioration, the cause and effect
relationship is not easily established. This research will
be exceedingly expensive.

Recognizing the relatively fixed amount of ground
and surface water supply, the increasing water needs
of the general population and industry, and the need
to reuse the available supply to satisfy future demands,
we can no longer afford to wait and see what happens.

The types of research mentioned are essential and
minimal. But this generation also bears a responsibility
for the health and well-being of future generations.
Realistically, answers to many of the currently identifiable research problems must be obtained quickly so
planners of our environment can formulate rational,
economical, and effective plans for the continued growth,
development, and survival of our society.

Synthetic chemicals. A most important achievement in
the past 3 decades has been the development of longer
lasting, more effective, and more toxic organic insecticides, a breakthrough the significance of which in terms
of human health has been of the highest order. It was
one of the most spectacular advances in improving man's
health the world has ever seen. But in terms of other

factors engendered by them, they became a mixed blessing.

It is impossible to say just how many hundreds of thousands of individuals have had the opportunity to live out their life-span or, at least, die of accident, chronic disease, or starvation rather than to die of malaria, typhus, or some other arthropod-borne disease, probably at an early age. The advent of these longer-lasting or more toxic compounds made possible the control of such commonplace diseases as shigellosis that, particularly in the neonatal phase of life, has produced so much misery and death. While resistant strains of arthropods were developed rather quickly and chemical control was not permanently effective, the few years of its effectiveness removed the apathy and the general acceptance of these diseases as routine risks in many parts of the world. Chemical control of flies, for example, is not as effective or as desirable as improved sanitation, but the latter had been hard to sell prior to the demonstrated benefits of flyless life through pesticides. The real impetus of the increasing use of pesticides, of course, has been the tremendous benefits they have brought in production of food and fiber.

We are now faced with the necessity—not merely the desirability—of evaluating the unanticipated effects of the host of chemicals which have been in widespread usage in ever-increasing amounts since World War II. These chemicals are not limited to pesticides—pesticides comprise only a small part of the problem. Many more chemicals are used as food additives, drugs, other household materials, and a complex of industrial effluents that get into the air, food, and water on which man depends. The enormity of the task is not only beyond our economic ability, but almost beyond comprehension, when we know that hundreds of these substances have the potential for interaction with one or more other substances. With limited talents and resources, problems must be selected that are representative of many others and attempts made at their solution.

An illustration is provided by the antibiotic griseofulvin. This pharmaceutical is administered by mouth over long periods to humans with certain fungal skin infec-tions. In the laboratory, griseofulvin produces a high incidence of liver tumors in mice at total doses far less than those used therapeutically in man. Meanwhile, a comparatively inert and nontoxic compound, piper-onyl butoxide (PB), used in agricultural pesticidal for-mulations to augment or synergize the effects of pyre-thrums has been developed. The action of PB appears to inhibit detoxifying enzymes in the insect; thus the insect becomes more sensitive to the insecticidal effects of pyrethrum. Piperonyl butoxide administered alone to infant mice was neither toxic nor carcinogenic. How-ever, when administered together with a variety of other agents at nontoxic levels, including griseofulvin, 3,4-benzpyrene (a polycyclic carcinogen widely distributed in the environment), and certain Freons (fluorocarbons with a wide range of domestic and industrial uses), the combination produced a marked synergistic toxicity; additionally, in the case of Freons, combined administra-tions with PB also produced liver tumors.

These results suggest the need to consider interactions, synergistic and otherwise, between unrelated and related agents when testing for effects of environmental pollut-ants.

One of the most serious problems that we are facing is not generally recognized, i.e., measuring the long-term hazards of popular materials. Pesticides, for instance, are usually evaluated to acute toxicity. The upper limit of tolerance is established in test animals by administration of high dose rates over short time periods. Often acute toxicity levels are far above exposures man would ever encounter. But we are beginning to recognize more and more chronic hazards—problems resulting from long-term exposure to low concentrations of these chemicals. Thus, acute or short-term hazards are no accurate meas-ure of chronic or long-term hazards. We must develop methods of evaluating both acute and chronic hazards before irreparable damage is done to mankind or his en-vironment.

It may prove to be true that some substances are car-cinogenic, irrespective of dosage, but since so many known carcinogens are widespread in our environment, we must develop means to assess our risks. The idea of identifying carcinogens by high dosages administered to exquisitely sensitive animals was predicated on the hope that if such substances were so identified, perhaps man could avoid them and, further, that unless we could identify a no-effect level we should assume that none existed. Carcinogenic food additives and pesticides could be banned and naturally occurring carcinogens could be avoided. This idea developed before our analytical capabilities had included a capacity to recognize the wide distribution of very small amounts of known carcinogens.

We now know we cannot avoid them entirely. We need to assess the relative risks entailed to guide our future actions. Unless we do so, we are heading for an early demise of the industrial research and development of pesticides, food additives, and other useful chemical compounds. The already high cost of such research is being driven to new heights. Many companies are sharply curtailing development of new products simply because their research resources are wholly engaged in defensive research to prove the safety of established products. This is the dilemma—not only benefit versus cost but also benefit versus risk. These companies are committed to the premise that products must be as safe as is possible, that risks should be minimal. They firmly believe that a long record of wide usage has proved the safety demanded.

The alternatives are not simple but they suggest that the Federal dollar must be spent either *a*) to develop and safety test new compounds (after which companies will vie with each other in efficient production), or *b*) to develop protocols for testing that are reliable and feasible for the companies to undertake or to contract for with commercial research laboratories. Otherwise, these extra costs will increase product prices markedly. Another approach might be legislation to permit industrial collaboration which otherwise could be viewed as industry-wide price fixing, outside the public interest. Science can study, but only the public can make such decisions.

A new approach to research must be developed. While

Federal laboratories have increased tremendously in scope and ability during the past 2 decades, and additional Federal funds have been devoted to the development of research potential and manpower in our universities and nonprofit research institutions, these growth curves have lagged behind the growth of research and development in industry. Today, this trend is being reversed. Even when allowances are made for the promotional aspects of the "D" in "R and D," industry has provided the bulk of the funds. In some segments of industry, however, such as aerospace, tax dollars support much of these industry expenditures.

With decreasing Federal support, it was hoped that industry, through product sales, would take over. But the Food and Drug Administration has the policy of requiring industry to pay the bill for testing—a cost passed on to the consumer, who is also the taxpayer. In this application, this policy appears sound. However, regulatory demands for more elaborate testing to insure safety are adding unprecedented costs to industry's research effort.

These alternatives are based on present methods of requiring industry expenditures for safety testing. But if a new factor could be introduced into safety testing, i.e., industry, government, and academic cooperation, if tax dollars could be used to develop, test, and evaluate protocols for establishing safety, the cost to both industry and the consumer who pays the taxes, directly or indirectly, could be reduced. This can be accomplished only if communications among industry scientists, regulatory scientists, and academic scientists are vastly improved and a concerted effort made to coordinate existing knowledge as well as discoveries derived from research.

At this moment, there is great need for reliable methods of extrapolating from animal experiments to man. While one could hardly expect this extrapolation to be either direct or simple, it would help tremendously if we could determine with repetitive reliability the dosage levels at which no chronic adverse effect can be expected in man. How this can be done is a subject of much discussion at present. From this discussion will emerge theoretical approaches to the solution. Federal funds are needed to test these approaches. Once reliable methods have been demonstrated, industry risk funds again will be expended. The probability of this will be vastly enhanced if the scientific community, government, industry, and academic, can become cooperatively involved.

This cooperation might well be regarded as the outstanding hoped-for breakthrough in the future since it is the essential ingredient to the success of developing safety standards. Now, no proof of safety exists; we only assume safety in absence of evidence to the contrary.

Meanwhile, it is reasonable to expect the following specific advances to take place:

a) continued development of biological pest control. Research on pest predators, radiation sterilization, and chemical sterilizants has increased the potential of control through means other than conventional pesticides, and thus reduced the risk that they offer to human health.

b) continued development of specific pesticides. The development of agents which have control effects on specific pests permits more precise use of pesticides and reduces the necessity for widespread coverage of agricultural areas with a general purpose agent irrespective of local essential requirements.

c) development of relatively safe, nonpersistent pesticides. The nonpersistent pesticides in current use are, for the most part, toxic to man in the form in which they are used. Fatalities are not uncommon in children who have access to material, and accidental poisonings occur in the applicators and other workers in recently treated crops. Safe forms of nonpersistent pesticides should be under development.

d) development of integrated chemical and biological control programs. The most effective and safest combinations of chemical and biological control agents for various crops are under investigation and should lead to optimal combinations in the next decade.

e) development of presently unavailable data on the more subtle and slowly produced effects of synthetic chemicals and pharmaceuticals individually and in combination on the basis of epidemiological retrospective and prospective studies of very large samples of humans.

f) development of more effective and time-compressing methods of laboratory animal assays of synthetic chemicals and pharmaceuticals.

The principal effect of the six developments just mentioned would be reduction of the chances for the initiation or the promotion of long-term degenerative disease processes such as carcinogenesis. To this should be added a corresponding reduction of the chances of teratogenesis and mutagenesis from the action of certain agricultural and household chemicals, as well as pharmaceuticals, on the human reproductive system.

SUMMARY AND CONCLUSIONS—CHEMICAL HAZARDS

The overall picture of significant environmental chemical pollutants and the disease states which either have been proved or are highly suspected to be related to them has been summarized in Table 3. It is noteworthy that these diseases head the list of causes of morbidity and death as well as of shortening of life expectancy. The following conclusions have been drawn with reference to chemical hazards to man:

1) Airborne pollutants possess greater potential for contributing to man's deteriorating health with age than do waterborne and foodborne contaminants together.

2) As a rule, pollutants express their effects only through interaction with other agents or with some preconditioning factor(s) within the host (a natural consequence of their extremely low ambient levels) be they infectious agents, trace-element deficiencies, or genetic defects of metabolism. Thus environmental pollutants must definitely be in-

TABLE 3. *Disease states for which evidence points to environmental pollutants as either direct or contributing causes*

Disease	Geographic Distribution		Relative Incidence Index[a]	Etiologic Pollutants and Associated Conditions	Direct	Contributing
	General	Localized				
Accelerated aging	+		High	Ozone and oxidant air pollutants	+	
Allergic asthma		+	High	Airborne denatured grain protein and other	+	+
Cardiovascular disease	+		High	"Hard" waters and hereditary tendency, Cr-deficiency states, CO(?)		+
Atherosclerotic heart disease		++				
Berylliosis		++	Very low	Airborne Be compounds	+	
Bronchitis		++	High	Acid gases, particulates, resp. infection, inclement climate		+
Cancer of the G.I. tract	+		Medium	Carcinogens in food, water, air and hereditary tendency		+
Cancer of the respiratory tract	+		Medium	Airborne carcinogens and hereditary tendency		+
Dental caries		+	Low	Se	++	
Emphysema	+		Medium	Airborne respiratory irritants and familial tendency	++	
Mesotheliomas	+		Low	Asbestos and associated trace metals and carcinogens (air, water) (other fibers?)	+	
Methemoglobinemia infant death		++	Low	Water-borne nitrates and nitrites	+	
Renal hypertension		++	Low	Cd in water, food and beverage in As and Se-low areas(?)		+

^a A composite index derived from an estimate of incidence, geographic extent and seriousness of effect. (From Stockinger (11).)

cluded among the multiple factors in the causality of chronic degenerative disease.

3) Acceleration of aging is the dominant characteristic of the effect of many of the top eight of the environmental pollutants. The imposition of such effects on the normal process of aging, whose complexities are only beginning to be understood, compounds the difficulties in either determining causal relationships or assessing the degree of contribution from environmental pollutants.

4) Finally, decidedly in the overall evaluation are the counteractants, the natural antagonists existing both in the environment and within the host. To measure only the pollutants without the counteractants can lead to an overestimation of the health problem. The homeostatic mechanisms leading to adaptation provide the balance to counteract the effects of pollutants at existing levels in the United States among the majority of the population. It is only when this balance is upset through predisposing disease or genetic fault (susceptibility) that environmental pollutants exert effects on man.

Major difficulties militating against the adequate solution of these environmentally induced disease problems include our inability to isolate the effects of any one chemical pollutant from others to which we are exposed, the long latent period for some cancers (e.g., 18 years for aromatic amine-induced bladder cancer), the longer latent period for genetic effects (e.g., more than one generation), our grossly inadequate base-line data due to the lack of national registration systems for cancer, birth defects, genetic defects, occupational disease, and the like, and the gross insensitivity of our animal testing procedures.

Realization of these concepts and also of the high total costs of human disease clearly indicates the need to strengthen anticipatory and preventive approaches, apart from improving our techniques for testing, recognition, and measurement.

What is required is an adequately supported interdisciplinary investigation to determine the distribution of the respective chemical hazards in the environment and their availability to and effects on plants, animals, and man; to consider ways and means of standardizing data collection and analysis and computer storage and retrieval; to establish avenues of communication and ways of disseminating information among the interdisciplinary groups; and to promote interdisciplinary national and international education in regard to the chemical environment and its effect on health and disease. The efforts of such an investigating group would provide an extremely important start toward the hoped for identification on a statistically sound basis, those correlations between geochemical and disease patterns that do, in fact, exist.

PHYSICAL HAZARDS TO MAN

Physical hazards in the environment probably represent the oldest recognized by man. The impact of various geological perturbations such as earthquakes, volcanic eruptions, floods, and tidal waves must have been dramatically evident. Yet, while natural disasters represent one of the major areas of threat to the existence of man, civilization and its resulting urbanization and industrialition has brought with it a number of physical threats that make the natural hazards appear minor. Three outstanding man-made threats to himself are radiation, noise, and accidents.

Radiation

Man has always been exposed to radiant energy, from the sun and from minerals. The extent of the role natural radiation has played in the evolution of man is a conjecture. The discovery of artificially produced X-rays by Roentgen in 1895, and successful nuclear fission in the 1940's, significantly added to the problem of radiation.

The new sources of radiant energy range from large-scale applications of nuclear energy, especially for electric-power generation, through lasers and microwave technology in industry, to the use of radionuclides and X-rays in the healing arts, the rapidly increasing use of microwaves by the communications industry, and in electronic equipment in the home. Our scientific knowledge and protection against this radiation is still at a very early stage. The extensive use of X-ray and other devices based on radiant energy has added appreciable exposure loads to large numbers of patients. Between one-third and one-half of all critical medical decisions are dependent on radiological information. However, only one-sixth of the world's population has access to modern radiology. As a result, throughout the world, including the United States, a large proportion of diagnostic X-ray films are inferior, uncertainly exposed, insufficiently collimated, poorly developed, and, therefore, difficult to interpret.

There are great disparities in the amount of radiation exposure used for comparable procedures and in the levels of genetic radiation doses. Even one X-ray during pregnancy has been found to increase significantly the chances of a child developing cancer during the first 10 years of life. This risk is greatest during the first trimester but exists throughout the pregnancy of the mother. Thus, in a study at the University of Oxford of 15,298 children born between 1943 and 1965, half had died of malignancies before the age of 10 years. The other half served as matched controls. Almost twice as many of the children who developed cancer had been exposed to X-rays before birth as had the children in the control group. If only one X-ray had been taken the increased cancer risk was 1.26 to 1. If five films had been taken, the increased risk was more than double, 2.24 to 2. X-Rays taken during the first trimester of pregnancy led to more than an eightfold increase in risk of childhood cancer.

A recent World Health Organization Expert Committee found that "the types of cancer in man that are directly or indirectly due to extrinsic factors are thought to account for a large percent of the total cancer incidence." It concluded, "Therefore, the majority of human cancers are potentially preventable."

The rapidly expanding use of ionizing and nonionizing radiation for weapons manufacture, power development, industrial uses, communications, and other purposes introduces to the air, water, and land a pollutant awesome in its potentials and implications. Merely the problem of safe disposal of the large amounts of radioactive wastes is beginning to appear overwhelming. Added to this are the dangers of leakage from stored materials, of radioactivity transmitted to cooling waters, and the problem of thermal pollution of streams and lakes, possibly eventually of even the ocean itself.

Of the total radiation exposure to which people in this country are subject, 45% is from natural sources, such as from minerals and from the sky (cosmic); 55% is from man-made sources, most of which (45% of the total) is from medical equipment. Industrial and occupational sources account for 7% of the total, and television screens, luminous clock-watch dials, and fallout plus fission testing, account for 1% each. It is important to remember that radiation cannot be directly detected by the senses and that the effects of radiation are irreversible. There is no immunity against radiation; parts of the body escaping damage from one exposure do not have an increased tolerance for future exposures.

Radioactive minerals provide real radiation hazards for the men who mine them. Among a group of 907 white uranium miners with more than 3 years of underground experience, death rates were 17.8 times the normal rate from heart disease, 5 times the normal from respiratory cancer, and 4.5 times normal from nonautomotive accidents; of 6,000 men who have been uranium miners, it is estimated that 600 to 1,100 will die of lung cancer in the next 20 years.

Two outstanding achievements of recent decades in the field of radiological health have been the development of an understanding of the genetic effects of ionizing radiation and the development of a logical basis for standards in radiation protection. Some of the genetic changes that can be effected by radiations and other hazardous agents are discussed in another chapter. The development of standards in radiation protection has resulted from the cooperative efforts of many people and a number of different organizations.

In radiation standards development perhaps the best references are found in the published reports of the Federal Radiation Council, the National Committee on Radiation Protection and Measurements, the International Commission on Radiological Protection, the United Nations Scientific Committee on Atomic Radiation, and the standards hearings of the Joint Commission on Atomic Energy.

Research on the genetic effects of radiation provided an experimental basis for determining the limits of radiation exposure to large groups of people and for improving national and international standards. A nuclear test ban treaty was stimulated by the data accumulated. While human health was not improved in the sense of curing radiation sickness, it was improved in a preventive sense. Undoubtedly millions of people are enjoying better health today because of the improved standards that have been developed.

These research developments and higher standards have, from a preventive standpoint, led to a better basis of control in the nuclear industry, in military applications of nuclear energy, and in the medical use of radiation, as in X-ray machines. For the most part, the scientific community and the government agencies involved deserve great credit and recognition for meeting this important problem in a most orderly way. Yet, even more could have been achieved more effectively if organized medicine had taken a more constructive approach toward a critical review of its practices and had encouraged a greater effort in establishing medical X-ray standards and techniques for compliance with them. Many physicians provided real leadership in the effort but more valuable human data could have been obtained if physicians had participated to a greater extent. Had it been possible earlier to have openly and freely discussed such things as plutonium, tritium, iodine, and krypton from the standpoint of radioactivity, earlier recognition could have been given to the problems involved in the use of these and other radioactive substances in the medical and bioscientific areas.

Although there continues to be some criticism of the

established standards most of the knowledgeable scientists and organizations find them acceptable, while continuing periodically to reevaluate them relative to necessary improvements. The adequacy of ionizing radiation standards could be made better in the future if additional information were available on the long-term effects of the radiation on animals and man, obtained through long-term animal studies and epidemiological studies of human populations. Such studies would require long-term investments, best assured as a national commitment to this purpose.

Because of the extensive work already done in the field of ionizing radiation, dramatic breakthroughs in this area are not very likely to occur in the near future. It is anticipated, rather, that painstaking research will produce new information in regard to such things as the genetic effects of radiation, the problem of radiation exposure during pregnancy, and radiation problems related to tritium. It is more likely that breakthrough types of advancement may occur in the field of nonionizing radiation, which has been investigated to a less extent, or perhaps in the area of sonic radiation (ultrasonics) and its biological effects. Perhaps the achievements to date in the field of ionization radiation and the ionizing radiation standards will serve as a stimulus and a model for advances in the more slowly developing field of nonionizing radiation.

A system should be established which makes it mandatory to allow scientists within the government or working under government contract to publish their findings without restriction in appropriate scientific journals irrespective of administrative clearance (other than national security). This would help the scientists and would provide a forum for discussion in the scientific community. It would also protect those administratively responsible because it would minimize the import that scientific papers represent on agency position. The scientific efforts of the personnel of an agency should not become automatic determinants in the development of policy, and administrative considerations should not be allowed to prevent the results of scientific research efforts from reaching the appropriate scientific community.

Another of man's environmental physical hazards is noise. This immediately gives rise to perplexing questions: What is noise? Can it actually produce ill effects? If so, how, under what circumstances, and with what other factors? Are any ill effects reversible? How might they be prevented? These question indicate noise to be the least understood of the environmental hazards. The only way to delineate the noise problem of a community is to note the sociological changes that might occur. If this could be done, then it might be possible to judge whether noise is totally, partially, or not at all responsible for any changes. Noise is more difficult to deal with than most other nuisances or environmental factors because it is partly subjective. In any society, each individual must necessarily accept a certain amount of annoyance, inconvenience, and interference. The essential question is how much and at what point may actual harm become a possibility.

There are three measurements of noise as a hazard to health: its intensity, its frequency, and the length of exposure. Depending on these, the environmental circumstances and the individuals involved, the effects of noise fall into four general categories: annoyance, disruption of activity, loss of hearing, and physical or mental deterioration. Annoyance is the most widespread response to noise. Some claim that annoyance is not related to health, but the issue is academic since noise abatement can be justified from either standpoint. An annoyance condition may aggravate existing physical disorders. Noise which disrupts sleep can lessen the body's resistance to disease or physical stress, and noxious noise generally disturbs one's feelings of well-being. Excessive sound can lead to somatic manifestations such as stomach problems, including ulcers, and allergies such as hives. In certain instances excessive noise is thought to aggravate mental illness. Evidence has also been found that exposure to certain types of noise causes constriction of the blood vessels near the skin surface, an effect that does not disappear with adaptation to the noise.

Noise can cause the blurring or masking of speech and

other wanted sounds. Thus, noise from machinery may interfere with instructions to a worker. A number of hospitals, especially near airports, have noted an apparent effect on the ability of patients to convalesce. In the field of education noise disrupts attention and hinders concentrated mental effort.

The greatest physiological effect of noise is hearing loss, temporary or permanent. Temporary impairment of hearing, called auditory fatigue, occurs after short exposure to intense noise. Exposure to a continuous high level of sound with inadequate recovery time between exposures may lead to permanent hearing damage.

In terms of the measurements of intensity, frequency, and length of exposure, their effect on hearing loss is as follows: intensity—exposure to sound pressure levels of over 80 decibels is hazardous, but there appears to be no permanent hearing hazard for levels below 80 decibels. Frequency—the inner ear is more susceptible to damage at middle and especially higher frequencies in the audible range than at low frequencies, and damage increases with increased exposure time. In addition, a fourth factor is the susceptibility of an individual's inner ear to noise-induced hearing loss. Individuals vary in this regard, but about 3% of the general population may be classified as highly susceptible. A survey between 1959 and 1961 by the Public Health Service of the population of the United States showed that the rate of hearing impairment per 1,000 persons was 7.6 for those under 25 years of age, 22.2 for those aged 25 to 44 years, and 51.2 for those between 45 and 64 years. Continued exposure to loud noise was believed to be the major cause of the increase as years of life passed. This is in agreement with the estimate that about 18 million Americans suffer total or partial deafness and that among working males two-thirds of the hearing loss is caused by noise. The progressive loss of hearing as age progresses is known as presbycusis.

Research on hearing acuity has been carried out among the Nilitic tribes in the Sudanese desert. These people live in probably the most noise-free area on earth. They have no musical instruments and apparently do not even sing. Remarkable hearing acuity has been observed, exemplified by their ability to communicate with a normal speaking voice over long distances and a sustained hearing acuity not only among the adults but also in the aged.

However, in the field of noise as an environmental health hazard, the progress made is the result of patient studies that often appear pedestrian but which are necessary if realistic damage–risk criteria are to be established. To assess the physiological effects of noise, one must either find a population of persons whose necessary noise exposures can be measured in detail or expose experimental animals under controlled conditions. Because of individual differences in susceptibility to damage by noise, either course involves a large number of subjects and is, therefore, expensive.

Although no breakthroughs can be indicated, progress has been steady. It is now known how much steady 8-hour/day, 5-day/week noise can be endured with negligible risk to hearing. We are on the way toward determining equally reliable damage–risk criteria for intermittent noise exposure and for impulsive noises such as gunfire. The quantification of the relations between the physical and temporal characteristics of intense sound and the degree of both temporary and permanent losses in hearing thus induced has been a significant accomplishment in the past 30 years. Now there is some agreement that hazards to the hearing from exposure to noise can be predicted with confidence. As a result it has been possible to set tolerable limits of noise exposure in both industry and in the community. Noise-induced deafness is a pervasive disease and social handicap that is now suffered by millions.

Once it could be shown that a level of 80 decibels or higher was hazardous, it was easier to convince workers that the use of ear protection devices was good sense. This, of course, is more of an educational achievement than a research one, but there now are young drop-forge operators, riveters, jet mechanics, and policemen who still have normal hearing because they use these ear protections.

Perhaps the most significant results achieved in the past 3 decades have been the *negative* ones—that is, studies that show that some of the effects attributed to noise on an anecdotal basis are only folklore. Research on the effects of noise on sleep and on the neurovegetative system continue, and should continue, even though results are ambiguous. Study of the physiological changes in the cochlea associated with noise damage should also continue, even though there seems to be little hope that we will ever uncover any sort of ameliorative agent that will reduce significantly or partially restore the damage done by noise.

In the future, some of the more subtle effects of noise on mental and physiological stress may be discovered and measured. This would permit the specification of tolerable limits and the exercise of noise control. Present research data are very controversial and need clarification. *1*) How much noise energy is tolerable as a daily dosage if the exposure is intermittent instead of continuous? *2*) Are very young or very old people more susceptible than others to damage from noise? *3*) In the case of persons with noise-induced hearing losses so severe as to cause trouble understanding ordinary conversation, how can the information in the speech signal be recoded so that understanding is restored? *4*) Is hearing loss, as measured by changes in auditory threshold, really an adequate measure of the damage actually done by noise? (Recent animal research has indicated that extensive irreversible damage to the hair cells of the cochlea may occur without affecting threshold significantly; if this is confirmed, it could be considered a genuine breakthrough.) *5*) Is an ear that has already been moderately damaged more susceptible to further insult than an unsullied one? *6*) Are there any long-range consequences of repeated autonomic excitation by noise?

In addition to progress in the acquisition of physiological and anatomical knowledge, there is need, of course, for significantly improved technology in noise abatement with reference to heavy industry, housing, streets, highways, and planes. This appears to be coming to a head rapidly as a result of the burgeoning aerospace industry and its present and potential effects on the environment of man and other creatures.

Injury

With the decline in mortality rates from communicable diseases, the relative importance of injury as a cause of death has increased. Accidental injury, suicide, and homicide rank 4th, 12th, and 14th, respectively, among the leading causes of death. The 10-year trend (1958–1967) shown in Table 4 makes it clear that little recent progress has been made in reducing the hazards.

Although morbidity data have no real meaning with respect to homicide and suicide, the attack rate for accidental injuries is approximately 25%—exceeding the common cold in frequency. Some 52,000,000 annual injuries require medical care or at least 1 day of restricted activity. Injury victims require more general hospital beds than any other class of patient (20,000,000 bed-days/year) and require 80,000 man years of professional care.

The sophisticated statistics of the injury problem are not matched by the availability of control measures. Research on behavior, both aberrant and delinquent, has not yet provided enough information to move confidently with preventive programs. Research on injury epidemiology, agent modifications, and environmental adaptation, coupled with safety education, has produced some tangible benefits for injury control.

A most important development has been in precise methods of analysis of the causes of accidents. The epidemiological and biostatistical methods used in the study and control of disease have been applied to accidents. We now get valuable information on the major categories and specific kinds of accidents and injuries, those experiencing them, conditions of occurrence and the interrelationship between persons, environmental factors and specific agents producing the injury. This information is useful in designing methods to reduce accidents.

Another advance has been the development in the field of human factors engineering (ergonomics), the major

CHAPTER VII

TABLE 4. *Death rates in United States from accidental injury, suicide and homicide between 1958 and 1967*

Cause of Death	Death Rate									
	1967	1966	1965	1964	1963	1962	1961	1960	1959	1958
Accidental injury	57.2	58.0	55.7	54.3	53.4	52.3	50.4	52.3	52.2	52.3
Suicide	10.8	10.9	11.1	10.8	11.0	10.9	10.4	10.6	10.6	10.7
Homicide	6.8	5.9	5.5	5.1	4.9	4.8	4.7	4.7	4.6	4.5

Source: Vital Statistics of the United States 1967.

objective of which is to reduce accidents and injury by better integration between man and the equipment he uses, paying attention to biological and psychological as well as physical factors. Originally developed in relationship to safer design of complex military and aerospace equipment, the ergonomic principles have been applied increasingly to civilian transportation and industrial operations.

Advances have been made in the development of greater protection of an individual from injury by redesigning equipment, such as machine-tool guarding, the padding of automobile instrument panels, recessing knobs and strengthening door locks on cars. Air-bag restraints are now being developed. However, illustrating the great need for behavioral research, only one in three drivers actually uses seat belts even where the law requires them.

Over the past 30 years there was a gradual improvement (decrease) in accidental death rates on a population basis. But improvements can be definitely cited only in the rates of injury in industrial work situations, and here the improvements seem to be getting less. The outstanding exception to improvement has been in the death rate due to motor vehicle accidents. The death rate remained the same for 20 years until 1960, but during the last 10 years it has increased possibly due to increased exposure to hazard and inadequate controls.

Much useful information on fatal and nonfatal accidents has been provided by the National Health Survey. Home accidents have been found to result in about half as many deaths as motor vehicle accidents, but they account for more than twice as many bed-disabling injuries and about seven times as many less severe nonfatal injuries.

Significant advances in accident prevention and injury control have been made by the application of biological and human factors information in air transportation and military flying. This information has been obtained by studies on human factors in relation to equipment design, from research on flight physiology and aerospace medicine, from the development of criteria for the medical and psychological selection of flying personnel and for maintaining their health and efficiency, and from especially intensive and extensive analyses, including autopsies, of accidental occurrences. The Armed Services schools of aviation medicine and the universities have helped to train the personnel for this work. No corresponding trained group exists for the control and treatment of highway injuries.

In automobile accidents the impact of the steering wheel and column is a major source of injury and death. A very substantial reduction in driver injury has been achieved by designing steering wheel rims and spokes so they would deform rather than break under impact and by using larger steering wheel hubs that would spread the impact forces. Another important development has been the seat belt, developed when studies showed that ejection from the car was a frequent factor in producing death or serious injuries. Seat belts reduce injuries and deaths, yet efforts to induce the majority of car occupants to use them have been unsuccessful. More recently combined lap–and–shoulder belts have been developed to both prevent ejection and reduce injuries inside the car, and they have proved very effective. In Sweden, in 28,000 accidents analyzed, no deaths of lap–shoulder belt wearers occurred at impact speeds less than 60 miles per hour, while fatalities did occur among non-wearers beginning at a 26 mile per hour impact speed.

In the United States, educational programs have helped reduce other types of injury by changing human behavior. In one Arkansas county within 1 year the incidence of burn injuries requiring hospitalization was cut to half of the 5-year average. The injuries were caused by improper wiring in the home, misuse of petroleum products, and proximity of flammable materials to heating units. In one county in South Carolina, hospitalization for poisoning accidents was reduced by 29% as a result of a 3-year educational project for parents of young children. In a Philadelphia area, a small-group educational program was carried on for a year relative to injuries occurring to members of the several groups or their families. In 1 year inpatient admissions to the hospital had decreased by 17%, and admissions from the areas in which the members of the discussion groups lived decreased by 26%. Some studies in the circumstances relating to injuries from falls of older persons, accidents with a high death rate, have provided helpful information relative to the severity of skeletal degeneration (osteoporosis) in these people, frequency of bone fractures, some of which were found to have preceded the accidental fall, and the like. Other studies have indicated that fluoridation of water supplies may retard the skeletal degeneration and perhaps decrease the frequency of bone fracture injuries in the elderly. A Swedish study showed that the rate of fractures was much higher in postmenopausal women than in men of comparable age.

Accidents lead all diseases as a cause of death between the ages of 1 and 37 years, yet research on the causes of accidental death has been relatively unsophisticated. Opportunities for the necessary professional training for accident research have been very limited. In 1969, only two of ten schools of public health offered training for this area. Nor have sustained funding patterns for non-transport accident research and control been developed. Currently only 1 dollar goes for research on accidents as compared with 300 dollars for medical research. There is no basic reporting system to provide data for the evaluation of community programs; systems for epidemiological intelligence or surveillance have not been developed on a geographically balanced basis. No central repository exists for information on injuries and rapid retrieval systems for data are only in development stages.

A significant reduction in the total number of accidents in the near future is unlikely but recent developments may eventually bring improvements. New concepts from the fields of biostatistics, medicine, engineering, and psychology are providing impetus for experimental studies and the design of safety programs. Work is underway to determine injury thresholds of the human body relative to impact forces, data important to the design of protective devices. Neurosurgeons have studied head injuries from a variety of high impact accidents. Public interest has increased somewhat and there has been national legislative activity such as the highway safety acts of 1966, legislation relating to fabric flammability and other hazards to child safety, product safety, and consumer protection; these can provide impetus for increased attention to accident potentials in private and industrial sectors. Health personnel and the medical profession seek to improve emergency medical services. But there is an urgent need for a strong centralized national accident prevention program to *1*) conduct research, *2*) provide technical assistance to states and communities, *3*) train personnel, and *4*) supply financial support. The savings could, in the long run, be enormous. An effective prevention and control program for accidents could result in *1*) an annual prevention of 30,000 fatal injuries, *2*) the saving of 60,000 people annually through prompt medical care, *3*) the prevention of 10,000,000 accidental injuries each year, *4*) the annual saving of 2,000,000 hospital bed-days now needed for accident victims, *5*) the annual savings of 8,000 manyears of medical care services, and *6*) an annual 3 billion dollar reduction in direct costs to accident victims. Accidental death and disability truly represent the "neglected disease" of modern industrialized society in this country.

SUMMARY AND CONCLUSIONS—PHYSICAL HAZARDS

These are but some of the environmental hazards with which modern man must contend. They have given rise to the rapidly growing field of Environmental Medicine which applies the principles and knowledge of biological and epidemiological research to an understanding of a complex of physical, physiological, and psychological disturbances. The results of these disturbances now constitute the major portion of accident conditions which come to the attention of the medical practitioner, unfortunately too often too late.

Patterns of morbidity and mortality are changing. Evidence from data obtained from past efforts may not be relevant to future events. People, their hazards, and their diseases are changing. The illnesses that will be experienced within a few decades will not be the same as those of today, much less of yesterday. Environmental factors certainly enter into their causation, but to what extent and in what manner is still largely to be determined, and only by means of research.

Environmentally induced chronic diseases are now known to have certain characteristics. *1*) Typically they result from low-level exposures, insufficient to produce acute reactions. *2*) They tend to develop insidiously over long periods of time. *3*) By the time they are clinically recognizable, the conditions are often irreversible. Emphysema, asbestosis, some cancers, teratogenic anomalies, and environmentally induced hearing loss are good examples.

Several years ago the direct health effects of environmental hazards on man were summarized by the President's Science Advisory Committee. Five categories of increasing severity were listed as follows: *1*) annoyance, irritation, and inconvenience, which while certainly real, produce effects that are uncertain and almost impossible to measure; *2*) physiological effects of unknown clinical significance, which occur on a transient basis and the cumulative effect of which is quite uncertain; *3*) worsening of existing diseases or disability and increasing the general level of sickness, the determination or evaluation of which is quite open; *4*) effects of a general increase in the death rate, again quite open; and *5*) the initiation of specific progressive disease, an area also with many questions unanswered.

The general conclusion was that the human effects of environmental hazards were very poorly understood but, because of the obvious importance of the effects, delay in intensive and extensive study could not be avoided.

In any consideration of most needed research in this field, four fundamental characteristics of environmental hazards stand out. *1*) They tend to be ubiquitous. *2*) They tend to be multifaceted as to sources and effects. *3*) They are generally insidious in their action time frame. *4*) They often act in concert and may potentiate one another.

Despite these characteristics it has been customary for science, industry, and government to study and deal with environmental hazards separately. For example, is mercury hazardous, in what physiological way, at what levels, and with what effects? The same question is asked about lead, pesticides, individual pharmaceuticals, sulfur oxides, and so on, each separately, each usually in terms of acute effects, and each in terms of toxicity alone, ignoring possible carcinogenic, teratogenic, mutagenic, and psychic effects. This approach has led to the fallacious tendency to decide that X is the toxic or dangerous level of a particular substance or hazard and that in order to be conservative perhaps one-tenth of that amount should be designated as the minimum toxic level. This ignores the fact that humans are not exposed to one substance alone but to an extensive and cumulative spectrum of hazards, each of which might even be minimal but the total burden or effect of which might be critical or overwhelming.

From an overall viewpoint, present efforts for the study of environmental hazards to modern man may be considered to fall into these categories: *1*) the search for and application of new experimental and clinical methods of defining health risks from known or potential environmental contaminants and hazards; *2*) more intensive studies of synergistic potentials of the innumerable combinations of environmental hazards with particular consideration of the development of toxicity, carcinogenicity, tetragenicity, mutagenicity, physical injuries, and psychic stress; and *3*) the categorization of physiological and biochemical hazards in terms of human population and risk.

Obviously any sound approaches to eventual protective measures must be based on the results of such essentially biological research.

SELECTED ADDITIONAL READING

General

1. *Restoring the Quality of the Environment*. Report of the Environmental Pollution Panel, President's Science Advisory Committee, The White House, November 1965.

2. *Environmental Pollution. A Challenge to Science and Technology.* Report of the Subcommittee on Science, Research, and Development to the Committee on Science and Astronautics, U.S. House of Representatives, 89th Congress, 2nd Session. Serial S, Washington, D.C. 1966.

3. *A Strategy for a Livable Environment.* Report of the Task Force on Environmental Health and Related Problems, U.S. Dept. of Health, Education, and Welfare, Washington, D.C., June 1967.

4. *Cleaning Our Environment: The Chemical Basis for Action.* A Report by the American Chemical Society, Washington, D.C., September 1969.

5. *Man's Health and the Environment—Some Research Needs.* Report of the Task Force on Research Planning in Environmental Health Science, U.S. Department of Health, Education, and Welfare, Washington, D.C., March 1970.

Biological Hazards to Man

6. STARR, M. P. (editor). *Global Impacts of Applied Microbiology.* New York: Wiley, 1964.

7. Communicable Disease Center, Annual Supplements, Morbidity and Mortality Weekly Reports, Summaries for 1951, 1960, and 1969.

8. MAY, J. M. (editor). *Studies in Disease Ecology.* New York: Hafner, 1961.

9. SARTWELL, P. E. (editor). *Maxey-Rosenau Preventive Medicine and Public Health.* New York: Appleton-Century-Crofts, 1965.

10. *National Center for Health Statistics—Vital Statistics Rates in the United States 1940–1960.* U.S. Department of Health, Education, and Welfare. U.S. Government Printing Office, 1968.

Chemical Hazards to Man

General

11. STOCKINGER, H. E. The spectre of today's environmental pollution. *Am. Ind. Hyg. Assoc. J.* 30: 195, 1969.

12. EPSTEIN, S. S. Control of chemical pollutants. *Nature* Nov. 27, 1970.

13. SINGER, S. F. (editor). *Global Effects of Environmental Pollution.* New York: Springer-Verlag, 1970.

14. *Clinical Handbook of Economic Poisons.* Washington, D.C.: U.S. Govt. Printing Office, 1963.

Naturally occurring chemicals

15. HEMPHILL, D. D. *Trace Substances in Environmental Health.* Columbia: U. of Missouri Press, 1969.

16. DIXON, J. R., D. B. LOWE, D. E. RICHARDS, L. J. CRALLEY AND H. E. STOCKINGER. The role of trace metals in chemical carcinogenesis—asbestos cancers. *Cancer Res.* 30: 1068–1074, April 1970.
17. SELIKOFF, I. J., E. C. HAMMOND AND J. CHURG. Asbestos exposure, smoking and neoplasia. *J. Am. Med. Assoc.* 204: 106–112, 1968.

Natural fungal or plant toxins

18. WOGAN, G. N. (editor). *Mycotoxins in Foodstuffs.* Cambridge: Mass. Inst. of Tech. Press, 1964.
19. RUSSELL, F. E., AND P. R. SAUNDERS (editors). *Animal Toxins.* New York: Pergamon, 1967.

Complex organic and inorganic chemical mixtures

20. GOLDSMITH, J. R. Effects of air pollution on human health. In: *Air Pollution* (2nd ed.), edited by A. C. Stern. New York: Academic, 1968.
21. ANDERSON, D. O. The effects of air contamination on health. *Can. Med. Assoc. J.* 97: 528, 585, 802, 1967.
22. MORROW, P. E. Some physical and physiological factors controlling the fate of inhaled substances. *Health Phys.* 2: 366, 1960.
23. KUSCHNER, M. The causes of lung cancer. *Am. Rev. Respirat. Diseases* 98: 573–590, 1968.
24. Toxicologic and epidemiologic bases for air quality criteria (special issue). *J. Air Pollution Control Assoc.* 19: 629–732, 1969.
25. SHUBIK, P., D. B. CLAYSON AND B. TERRACINI (editors). *The Quantification of Environmental Carcinogens.* Geneva: Intern. Union Against Cancer, 1970, vol. 4.
26. HARTWELL, J. L. (1951); J. L. HARTWELL AND P. SHUBIK (1957); J. PETERS (1969). *Survey of Compounds Which Have Been Tested for Carcinogenic Activity.* Public Health Service Publ. no. 149.
27. BOREN, H. G. Pathobiology of air pollutants. *Environmental Res.* 1: 178–197, 1967.
28. HANNA, M. G., JR., P. NETTESHEIM AND J. R. GILBERT (editors). *Inhalation Carcinogenesis.* Proc. Biol. Div. Conf., Oak Ridge National Laboratory, 1970.
29. EPSTEIN, S. S., AND H. SHAFNER. Chemical mutagens in the human environment. *Nature* 219: 385, 1968.

Synthetic chemicals

30. MOSER, R. H. *Diseases of Medical Progress* (3rd ed.). Springfield, Ill.: Thomas, 1971.
31. SELYE, H. Adaptive steroids—retrospect and prospect. *Perspectives Biol. Med.* 13: 343–363, 1970.

32. Report of the Secretary's Commission on Pesticides and Their Relationship to Environmental Health. U.S. Govt. Printing Office, 1969.
33. Report of the NCI/Bionetics study of pesticides. *J. Natl. Cancer Inst.* 42: 1101–1114, 1969.
34. CONNEY, A. H. Drug metabolism and therapeutics. *New Engl. J. Med.* 280: 653–672, 1969.
35. *Proceedings of Symposium on Microsomes and Drug Oxidations.* New York: Academic, 1969.
36. MANNERING, G. J. Significance of stimulation and inhibition of drug metabolism. In: *Selected Pharmacological Testing Methods,* edited by A. Burger. New York: Dekker, 1968.

Physical Hazards to Man

Radiation

37. INGRAHAM, S. C., J. G. TERRILL AND D. W. MOELLER. *Concepts of Radiological Health.* Public Health Service Publ. no. 336, Washington, D.C.
38. GITLIN, J. N., AND P. S. LAWRENCE. *Population Exposures to X-rays, U.S., 1964.* A Report on the Public Health Service X-ray Exposure Study. Public Health Service Publ. no. 1519, Washington, D.C.
39. *Report of the Medical X-ray Advisory Committee on Public Health Considerations in Medical Diagnostic Radiology.* Public Health Service, Washington, D.C., Oct. 1967.
40. *Radiation Exposure Overview: Diagnostic Dental X-rays and the Patient.* Public Health Service, Washington, D.C., July 1968.
41. CARPENTER, R. L., AND V. A. CLARK. Responses to Radiofrequency Radiation. In: *Environmental Biology,* edited by P. L. Altman and D. S. Dittmer. Bethesda, Md.: Fed. Am. Soc. Exptl. Biol., 1966.
42. *Radiation Bio-effects.* Summary Report, Public Health Service, Washington, D.C., 1968.
43. *Proceedings for the Symposium on Public Health Aspects of Peaceful Uses of Nuclear Explosives.* Clearinghouse for Federal Scientific and Technical Info., Springfield, Va., 1969.
44. MOORE, W. *Biological Aspects of Microwave Radiation–A Review of Hazards.* Public Health Service, Washington, D.C., July 1968.
45. *An Annotated Bibliography of Regulations, Standards and Guides for Microwaves, Ultraviolet Radiation, and Radiation from Lasers and Television Receivers.* Public Health Service, Washington, D.C., April 1968.
46. *A Study of the Radiological Health Aspects of Agents Modifying the Biological Effects of Radiation.* Bethesda, Md.: Fed. Am. Soc. Exptl. Biol., June 1969.

CHAPTER VII
47. BURNS, W. *Noise and Man*. London: John Murray, 1968.
48. *Proceedings of the Conference on Noise as a Public Health Hazard*, edited by W. Dixon Ward and James E. Fricke. Washington, D. C., June 13–14, 1968, Am. Speech & Hearing Assoc. Report 4, February 1969.
49. KRYTER, K. D. *The Effects of Noise on Man*. New York: Academic
50. *Noise:. An Occupational Hazard and Public Nuisance*. Chr. of World Health Organ., 20: 191, June 1969.
51. ROSEN, S., et al.: Presbycusis Study of a Relatively Noise-Free Population in the Sudan. *Ann. Otol. Rhinol. Laryngol*. 71: 727, 1962.
52. *Engineering Control, Industrial Noise Manual* (2nd ed.). Am. Ind. Hyg. Assoc., 1966.
53. COHEN, A. Noise effects on health, productivity and well-being. *Trans. N.Y. Acad. Sci*. 30: 910–918, 1968.
54. FARR, L. Medical consequences of environmental home noises. *J. Am. Med. Assoc*. 202: 171–174, 1967.

Injury

55. BACKETT, E. M. *Domestic Accidents*, Public Health Paper no. 26. Geneva, Switzerland: World Health Organization, 1965.
56. GALLAGHER, R. E., AND R. E. MARLAND. The Effect of Federal Legislation on Burn Injuries Involving Flammable Fabrics, Proc. First Annual Meeting, Am. Burn Assoc., Atlanta, Ga., March 1969.
57. GORDON, J. E. The epidemiology of accidents. *Am. J. Public Health*, 39: 504–515, 1949.
58. HADDON, W., E. A. SUCHMAN AND D. KLEIN. *Accident Research*. New York; Harper & Rowe, 1964.
59. ISKRANT, A. P., AND P. U. JOLIET. *Accidents and Homicide*. Am. Public Health Assn., Vital and Health Statistics Monograph, Cambridge: Harvard Univ. Press, 1968.
60. MARLAND, R. E. Injury epidemiology. *J. Safety Res*. 1: no. 3, Sept. 1969.
61. MARLAND, R. E., AND F. D. BURG.: The injury control program of the U. S. Public Health Service. *Pediatrics* 44: Suppl., 888–890, Nov. 1969.
62. McFARLAND, R. A. Injury—A major environmental problem. *Arch. Environ. Health* 19: 244–256, 1969.
63. McFARLAND, R. A., R. C. MOORE AND W. H. BERTRAND. *Human Variables in Motor Vehicle Accidents: A Review of the Literature*. Boston: Harvard School of Public Health, 1955.
64. NORMAN, L. G. *Road Traffic Accidents: Epidemiology, Control and Prevention*. Public Health Papers no. 12, Geneva, Switzerland: World Health Organization, 1962.
65. SUCHMAN, E. A. Cultural and social factors in accident occurrence and control. *J. Occupational Med*. 7: 487–492, 1965.

Chapter VIII

Marine sciences

CLARENCE P. IDYLL, Ph.D.

A MAJORITY OF THE WORLD'S POPULATION obtains at least half of its daily protein requirements from seafood and related products. Seafoods contribute less to the animal protein diet of most Americans because of the ready availability of meat, but they are still an important part of the preferred diet of many of our people. The lakes, streams, and rivers of the United States, and the seas adjacent to its coasts, play an important role in the welfare and economy of the country. These waters provide transportation, recreation, and avenues of waste disposal, as well as food. The aquatic environment of the United States, including the continental shelf to a depth of 100 fathoms, occupies about 17 % of the total area (4,271,074 square miles); of this, 15 % is seawater over the shelf and nearly 2 % is freshwater in lakes, streams, and rivers.

Research in aquatic sciences in the United States began seriously in the latter part of the 19th century, and has progressed most rapidly since 1940. Other studies became possible after the war because of the application of newly developed scientific theory and technology. Biological studies of the aquatic resources have contributed to human welfare by increasing the harvest of food from the ocean and freshwaters, by managing the resources and their environment so that food production may be sustained, and by promoting man's health and general welfare.

Primary organic production—the photosynthetic production of organic matter by green, chlorophyll-bearing plants—is the ultimate source of all life on earth. In the sea, the process of organic production is carried out mainly by small, usually microscopic, unicellular algae. These plants provide the basic food, directly or indirectly, for all other marine life, including species important to man as food.

In the early 1950's a technique was developed involving the use of radioactive isotopes as tracers that for the first time permitted the rapid, sensitive, and accurate measurement of the rate of organic productivity in the sea. This technique, which could be employed routinely on oceanographic expeditions, found widespread application to ocean exploration, and is now used extensively. As a result, man has a better knowledge of the total organic productivity of the ocean, as well as its seasonal and geographical variability.

The mechanisms and dynamics by which the photosynthetically produced organic matter is passed through the "food chains" or "food webs" of the sea, ultimately reaching species that are of direct value to man, are becoming better understood. We know enough to permit calculations of the potential food production from the seas. This potential is several times the present harvest. Whether man is able to realize this potential depends on a complex of technical, social, economic, and political problems. But some marine food resources are, or are about to be, overexploited. Basic studies in biological oceanography and marine ecology have pointed up the need for scientific management of the world's fisheries. In addition, it is essential to explore for new fishery resources.

Studies of biological productivity and the related physical and chemical oceanographic processes that limit and control productivity have made it possible to predict when and where to expect major concentrations of fish. Such information, together with knowledge of the magnitude and rate of renewal of the resources, permits their exploitation with maximum efficiency and safety for their preservation. Application of physical oceanography to the problems of fishery prediction is one of the more promising new approaches to the field of marine food resource management.

MARINE
SCIENCES

Knowledge of the chemical and physical mechanisms and processes that limit and control the production of organic matter in the sea make it possible to consider improving the ocean's productivity. Pilot-scale projects are being planned in which limited marine areas will be fertilized by pumping deep, nutrient-rich water to the surface. Hopefully this will increase the rate of photosynthetic production of the unicellular algae.

Of great importance for food production and protection of environmental quality are recent experiments using human wastes as nutrients for stimulating the growth of marine algae under controlled conditions, and feeding the unicellular plants to shellfish or other valuable food species. Wastes thus can be recycled for useful purposes. Such control and enhancement of basic biological productivity, coupled with an expanding ability to cultivate a variety of marine organisms under controlled conditions, give promise of a substantial increase in productivity of sea food.

The algae supply many basic nutrients, and are a rich source of many minerals (e.g., iodine, sodium, chlorine) and vitamins (e.g., thiamin, niacin, riboflavin, folic acid, ascorbic acid, and provitamin D). Some algae have a high protein content. For example, *Chlorella* can be cultured easily to contain 60% protein on a dry weight basis.

The large algae (seaweeds) have been used in cattle feeds (to increase butterfat content in milk), in chicken feeds (to provide for higher iodine content of eggs), and in fertilizer. The best known use of algae is in production of seaweed extractives, which are used extensively in laboratories as a growth medium (agar-agar) for cultivating microorganisms, and in products such as ice cream for consistency stabilization, in chocolate milk to provide body and suspend the cocoa, in eggnog as a stabilizer, and in milk-based cooked puddings as a solidifying agent. Seaweed extractives are also used extensively in dietetic foods to provide the body and texture normally supplied by sugar or starch.

CHAPTER VIII

The successful attainment and maintenance of the maximum yield of food from the sea depends in great degree on results of scientific research. Such research has several objectives: *1*) to locate, concentrate, and catch fish more quickly and less expensively; *2*) to improve and maintain the quality of the food product; *3*) to develop new products that will appeal to the consumer; *4*) to prevent overfishing; and *5*) to improve the economic position of the fisheries generally.

Fish and shellfish provide 10–13 % of world's consumption of animal protein. Most of this comes from the sea. Recently, the annual marine fishery harvest has been about 60 million metric tons, a weight nearly equal to that of the world production of meat and poultry. Most marine scientists agree that the ocean could produce more food, but it is not so clear how much more.

The potential world catch of fishery products is probably at least twice the present yield and may be considerably higher. The Food and Agriculture Organization of the United Nations suggests that upper limits on ocean fish production are conservatively 100 million metric tons for whales and marine fish; 2 million metric tons for shrimp, lobsters, and crabs; and 3 million metric tons for clams, oysters, mussels, squid, and octopuses. One method of increasing the present harvest is to use some of the many species that are currently not utilized. For example, over 50 % of the United States catch of marine animals consists of fewer than two dozen species. It may be that the natural production of some species of marine animals could be increased substantially by intensive cultivation by man.

RESOURCE ASSESSMENT

Systematic documentation of the marine animal resources inhabiting the continental shelf and slope adjacent to the United States was started in the latter part of the 1940's by the agency which is now the National Marine Fisheries Service of the U. S. Department of Commerce. Information has been collected on the bottom-

dwelling fish and shellfish communities in the northeast
Pacific Ocean from the Bering Sea to Baja California, and
on shrimp, scallops, lobsters, clams, and industrial fish
species along the Atlantic seaboard and in the Gulf of
Mexico. Seasonal distribution and abundance of a number of fish species that are major resources have been
documented and subsequently became the objective of
commercial fishing activities.

Fisheries which developed as a direct consequence of
this explorative research in the northeast Pacific include
the king crab fishery of the Bering Sea and the Gulf of
Alaska, the large deep-water (pandalid) shrimp fishery
of the Gulf of Alaska, the hake fishery in Puget Sound and
off the coasts of Oregon and Washington, the ocean perch
fishery off Oregon, Washington, California, and British
Columbia, the scallop fishery of the Gulf of Alaska, and
the deep-water flounder fishery off the coasts of Oregon,
Washington, and British Columbia. Along the Atlantic
seaboard, assessments conducted by U. S. scientists have
expanded the shrimp fishery in the Gulf of Mexico, aided
development of shrimp fisheries in Central and South
America, led to the development of the calico scallop
fishery and the major expansion of the industrial fishery
in the Gulf of Mexico, and contributed to the development of the clam fishery and the deep-sea lobster fishery
of the continental slope and the swordfish industry in
offshore waters.

Some indication of the value of explorative research
can be derived from the following figures: the Alaska
pandalid shrimp industry in the early 1950's was harvesting from 1 to 2 million pounds annually. In the 12 years
subsequent to assessment of activities in the northeast
Pacific, this fishery industry expanded rapidly, and is
now harvesting in the order of 60–70 million pounds.
Its annual value at the fisherman level is approximately
$3.5 million. The Alaska king crab fishery started follow-
ing assessment activities conducted in the Bering Sea and
Gulf of Alaska in the latter part of the 1940's. The annual
value of this fishery in recent years has exceeded $10
million. Shrimp are now harvested by vessels operating
in Central and South America as the result of explorative

research started in the early 1950's. In 1969, American flag shrimp vessels fishing out of Trinidad, Barbados, Guyana, Surinam, Nicaragua, and French Guiana harvested shrimp valued at over $32 million. This research has also provided jobs in Latin American countries and assisted economic growth in these countries.

When deep-water explorations were undertaken off the coasts of Oregon, Washington, and British Columbia, commercial trawling activities in these waters were limited to depths shallower than 100 fathoms. With the discovery of large concentrations of Pacific ocean perch and flounders on the continental slope, the commercial fishery rapidly expanded into deeper waters, and ocean perch became the most important species in the trawl fisheries in Oregon and Washington. Similarly, Dover sole, which was not exploited to any extent in the northern waters, has become one of the dominant species in the trawl landings. In addition to these species, major fisheries were also developed for Alaska scallops and sablefish on the continental shelf. The assessment of the hake resources in the mid-1960's resulted in development of a United States-based fishery in Puget Sound, and large-scale Soviet activities off the coasts of Oregon and Washington.

Major contributions have been made to both U.S. fisheries and the growth and development of foreign fisheries by explorations conducted by the United States. Foreign vessels harvest some 3 billion pounds of fish annually in areas surveyed by U.S. scientists. The scientific assessment also has included descriptions of the oceanographic features of the area, such as current patterns and temperature regimens. The studies of the International Pacific Halibut Commission pioneered the field of oceanic hydrographic studies in the North Pacific. A prominent seamount was discovered by fishery scientists for which there are now plans to establish a hydrographic laboratory and weather station.

The resource assessment studies have benefited the commercial fishing industry in other ways. A large midwater trawl developed for hake assessment is now used in the commercial trawl fishery of the West Coast. Similarly, a sorting net developed for shrimp sampling has contributed to conservation in that it releases undersized flounders and other fishes, meanwhile maintaining a high quality of shrimp harvested.

The assessment programs also have made important contributions to underdeveloped lands. The United Nations Food and Agriculture Organization and the USAID activities have used many United States experts to conduct resource assessment work in the Indian Ocean, South America, Africa, and elsewhere.

BEHAVIOR STUDIES

Fishermen have learned that the more knowledge they have of the habits of their prey, the more successful they will be in locating and capturing it. Answers to questions about the behavior of commercially important marine animals in relation to fishing techniques and tactics are being obtained from a combination of research and experience.

The information most important in predicting areas of abundance is on the responses of animals to temperature, salinity, and food gradients of the ocean, and the yearly and daily migrations that species make. Once fish are located, choices of gear and tactics are dictated by the animals' social behavior and swimming activity, and their responses to the gear itself.

An important role of behavior research relates to the unknown effects of pollution. Most of the world fishery resources spend a part of their lives in estuaries. To preserve fish and shellfish stocks it is essential to have knowledge of the behavioral and physiological responses of the animals to polluted and dammed water courses.

For example, when alarmed, many fishes can swim rapidly for short periods and can escape a net about to encircle or engulf them. The design and use of towed nets such as trawls depend greatly on the swimming abilities of the fish being sought. The predictive study of swimming speed has its roots in basic science. Accurate equations have been developed that predict "burst" or escape speeds of fish. There are differences in speed due to the shape and size of fish. To illustrate, contrast the requirements of a towed net to capture the large, powerful tuna which have burst speeds faster than 10 mph with gear for small flatfish whose burst speeds are no greater than 0.5 mph. Since towing speeds are usually less than 5 mph, it is apparent that a very large net would be needed to catch tuna. Studies provide computations relating escape speeds to the net diameter and towing speed. Design of gear must also take into account the behavioral consideration of how far the fish can sense and react to the net.

Information on behavior of fishes is required in the design of fishways. Many dams on the rivers of the Pacific Northwest have been provided with fishways to allow migrating salmon to pass upstream to spawning grounds. Early fishways were designed by rule-of-thumb, but the more efficient and economical structures have been designed on the basis of information from behavior studies. The Bonneville Fishery Laboratory, operated by the National Marine Fisheries Service and funded largely by the U.S. Corps of Engineers, has designed successful fishways for upstream migration. They have also designed collecting devices for young salmon migrating downstream, and have developed criteria for operation of turbines to reduce the mortality of the downstream migrants. Without this information the salmon runs in the streams of the Pacific Northwest could not be maintained at their present levels.

In Maine, behavior studies have led to the use of air-bubble curtains to guide herring schools to the point of capture. The reaction of herring to the air curtain appears to be primarily a response to visual stimuli. This technique has allowed the extension of fishing activities beyond the limits of depth and distance from shore imposed by the conventional passive methods of fishing with stop seines or weirs.

The purse-seine fishery for menhaden off the Atlantic and Gulf coasts of the United States has been made more efficient by utilizing the response of the fish to electricity. By fitting the anode of a pulsed direct-current circuit to the nozzle of the pump hose, the fish are attracted to the nozzle, and the efficiency of pumping the catch from

the net to the carrier vessel is greatly improved. Partly as
a result of this technique it has been possible to reduce the
crews from 22 to 10 men per boat, to increase the fishing
power, and to reduce the strain on the net. Electrical
gear has also been devised to improve the efficiency of
shrimp trawls. Some species of shrimp normally burrow
in the daytime, and an electrified shrimp trawl has been
designed and tested which can catch as many shrimp
during the day as at night. It is hoped in this way to extend fishing operations to a 24-hour basis.

These examples—some applied, others awaiting application—are instances of basic research on fish behavior
conducted in the past 2 or 3 decades. Far more basic
knowledge is required to improve the efficiency of location and capture of marine animals of economic importance.

VALUE OF PREDICTION OF FISHING SUCCESS

Much of the justification used for funding ocean research is based on the premise that knowledge of the sea
and its occupants is of practical value to military and
commercial interests. It is difficult to place a dollar value
on military use of this information, but commercially the
forecasting of fishing success has obvious potential.

The United States is the major world market for tuna,
shrimp, crab, lobster, salmon, groundfish, and fish reduction products. In the United States the value of all
processed fishery products produced from domestic and
imported raw material is now about 1.7 billion dollars
annually, a very substantial industry. Raw material acquisition has become a world wide task. Segments of the
U.S. fishing industry may be found in most of the fishing
nations of the world.

Due to the nature of the resources and present methods
of harvest, there are wide fluctuations in the abundance,
distribution, and availability of fish and shellfish stocks
from year to year and even within years. At the other end
of the chain, established food products, by nature, have

relatively stable markets. A major problem is to smooth out the oscillations so the processors and marketers may effect orderly merchandising.

The shrimp industry provides a practical example. The U.S. market utilizes more than one-third of the total world production of shrimp. Domestic production accounts for 45 % of U.S. needs, and a major share of this comes from the grounds in the Gulf of Mexico. These fisheries have been studied extensively during the past 15 years. Excellent statistical information on landings is available on a current basis. A historical study of supply–demand–price relationships in the United States reveals that the year–to–year fluctuations in the supply of shrimp from domestic sources are the key factors in determining trends in price of raw shrimp. Biological data show that: *1*) fishing effort for the three major species—brown, white and pink—in the Gulf of Mexico has reached a level where further increases in fishing effort have a minor effect on total landings; *2*) the major share of the landings are from the initial age group. Total stock mortality (from fishing and natural sources) is extremely high, and the recruits to the fishery are cropped off as they enter the fishery; *3*) fluctuations in year–class strength vary by a factor of two to six and are related largely to nearshore environmental changes. Forecasts of abundance from year to year can be integrated with market demand, price inventory, and import information to provide a sound planning base for fishing and processing operations.

For fishing vessel operations, the value of forecasting fish abundance is obvious. However, the present state of the art offers little practical assistance on a day to day basis, although some success has been obtained in special weather forecasts.

It is difficult to evaluate the dollar potential of weather forecasting to the seafood industry. However, a 5 % saving would be 85 million dollars annually. Probably less than 5 % of this amount is being gained as a result of present very limited forecasting operations.

Fishery scientists readily admit that present skills in predicting resource availability have not reached their full potential due to the lack of: *1)* current synoptic information on the physical properties of the air and ocean; *2)* basic research into the life histories of the animals under consideration; *3)* understanding and interest by fishery scientists in the application of their findings to the forecasting of fish landings; *4)* clear communication of the existing information to industry people in a usable form; and *5)* currency in the data that were transmitted to the fishing industry. Most studies that relate to the prediction of fish supplies utilize historical information and findings are published years later.

During the past 10 years, rapid advances have been made in certain of these areas. The sums spent on ocean research related to fisheries have increased. The results are encouraging. Synoptic data from satellites are now being processed and integrated with information obtained from surface sources to improve ocean and air data coverage on a real time basis. But information on the fish and shellfish stocks themselves is lagging. Worse yet is the lack of fundamental research linking the environment and fishing activities with stock abundance and distribution.

The reduction of ocean research budgets in recent years has accentuated the problem. New technology is being developed, but this is rapidly producing a gap between the hardware capabilities and the synoptic data being gathered and the software development (research) necessary to utilize the information intelligently. Steps must be taken to correct this deficiency or a major share of the projected gains will not be realized.

PRESERVATION OF FISHERY PRODUCTS

Fish are important in feeding the world's population, but this use is reduced because seafood products are extremely perishable. The availability of fish in inland areas requires inexpensive methods of preservation that keep the product wholesome. Historically salting, drying, smoking, and combinations of these methods have been used to retard decomposition. These methods are still in use in much of the world; they lend themselves to small operations, are simple and inexpensive.

In most industrial countries supplies of animal protein are abundant, and fish and shellfish are consumed because they are enjoyable components of balanced diets. In these countries the availability of refrigeration and the existence of highly developed distribution systems make fresh or frozen fishery products generally available. Consumers in these countries prefer their fish in these forms.

Refrigeration is a highly efficient means of controlling the growth of spoilage bacteria, but refrigeration methods still can be improved. Very recently research on spoilage control led to a major advance in the use of refrigerated seawater on board fishing vessels for holding fish at high-quality levels for extended periods of time at temperatures above the freezing point. The advantage of refrigerated seawater in the control of spoilage bacteria is that its freezing point is about 1 C lower than that of freshwater. This permits holding the fish in unfrozen state at a lower temperature than is possible with ice. The technique of saturating the seawater with carbon dioxide before adding fish to this refrigerant, and of keeping the seawater saturated during the entire holding period, permits an approximate doubling of the holding period. At the end of this holding period the quality of the product is significantly higher than that of the same products held for half that period of time in ice or in unmodified refrigerated seawater.

The bactericidal action of ionizing radiation has been known for more than half a century, but only recently has it been shown that meat can be sterilized by high-energy X-rays. In addition, penetrating electrons have been used as an experimental and demonstration tool for sterilizing a variety of raw food, and this led to a new method for preserving food. Ionizing radiation can be used to sterilize food, but the energy required to destroy all the bacteria present is so great as to alter the chemistry of the food product, producing undesirable flavors and odors. Research has shown that at lower radiation doses, preservation of fish and shellfish can be effected at a pasteurizing level (approximately 90% of the spoilage bacteria are destroyed) without altering the flavor or

odor of the product. Such a product requires refrigeration at temperatures of about 2–5 C; in this temperature
range fish and shellfish can be held for about 3 weeks
after irradiation—a time sufficient for shipment throughout the United States by surface transportation, and
for sale at the retail level as high-quality "fresh" products.

The control of spoilage in fishery products by radiation pasteurization does not destroy the deadly botulism
bacterium, *Clostridium botulinum* Type E. Fresh fishery
products sometimes carry small numbers of spores of
this pathogen, but they have never been known to cause
a case of botulism. As further protection of consumers,
recent studies on the mechanism of toxin formation have
shown that bacteriophages (viruses that attack bacteria
and are found almost universally in the vegetative cells
of *C. botulinum*), can be eliminated from the cells by
laboratory procedures, thereby causing the cells to lose
their ability to produce toxin. When the cells are rein-
fected with the bacteriophage, the cells become toxin
producers again. When bacteriophages from one type
of botulinum are added to bacteriophage-free cells of
a different type, the host of the "foreign" bacteriophage
produces toxin again but now the toxin produced is
characteristic of the type of botulinum from which the
bacteriophage was originally eliminated. This work has
significance for the entire food field, because *C. botulinum*
is a ubiquitous pathogen and is always a potential hazard
in a nonsterile food product that is stored after the normal bacterial flora have been altered by processing.

FISH PROTEIN CONCENTRATE

Research and development by private industry and
government during the past 15 years have resulted in
the development of other processes for making fish available where seafood is desired but is infrequently consumed because of its perishability or high cost. One of
these techniques is the production of concentrates of
fish protein. These are dehydrated and defatted fish,

usually reduced in particle size to a powder. Most developed processes are physiochemical in nature, but include some interesting biological processes. Research has shown that the protein content of the final product can be increased by the use of fish oil as a source of energy by microorganisms. As the organisms grow and break down the oil, they produce additional protein in the form of single-cell organisms. Other methods involve the enzymatic breakdown of fish protein to smaller molecules that retain the nutritive value of the original protein but simplify the removal of oil.

All processes are designed to produce a dry product that is highly resistant to bacterial and mold growth, acceptable as an ingredient of other foods, and retentive of the nutritional properties characteristic of animal protein. In the past 3 or 4 years plans for the commercial production of fish protein concentrate have become a reality in Canada and in Europe. The Federal Government has financed the construction of a demonstration fish protein concentrate plant in Aberdeen, Washington, that began operations in the Spring of 1971. The purpose of the plant is to demonstrate to industry and government the feasibility of economically converting little-used fish resources into shelf-stable, palatable, nutritious, concentrated protein and to provide facilities in which pilot-scale processes can be tested and modified to increase the efficiency of converting fish to protein concentrate.

Another interesting new technique for the control of spoilage organisms on freshly caught fish is being investigated currently. The principle involved is the absorption of heat by evaporating water and the liberation of heat by condensing water vapor. Fish are placed in a sealed chamber and subjected to a moderate decrease in pressure; moisture on the surfaces of the fish evaporates and quickly cools the surfaces by absorbing in the form of heat the energy required for evaporation; steam is then introduced into the chamber; the hot water vapor condenses on the surfaces of the cooled fish, giving up the latent heat of evaporation to the surfaces and, of course, to bacteria on the surfaces; after a few seconds a vacuum is applied again, and the fish are quickly cooled to refrigerator temperatures. The heat from the condensing steam is sufficient to destroy about 99 % of the vegetative cells on the fish, but the time allowed for heat absorption by the fish is so short as to obviate heat damage to the fish flesh. Taste-panel evaluations have shown that stored fish treated by this method are highly acceptable after conventionally iced fish are rejected.

MANAGEMENT OF RENEWABLE RESOURCES

Success in managing living marine resources and their fisheries depends as much on maintaining the yield of traditional resources as on developing unused resources. The period since 1945 has seen many advances that have contributed to fishery management. Major advances in the last 25 years include: *1*) clear demonstration that even in the ocean selective fishing can disturb the ecological balance, allowing other species to capture the energy released by reducing the abundance of exploited species, and this must be taken into account in research planning and analysis and in management practices; *2*) modern techniques for identifying stocks and races of fish, which permit identification of the origin of migratory species caught on the high seas; *3*) research leading to successful control of predators, especially sea lampreys in the Great Lakes, and boring snails and other predators of oysters and clams; and *4*) improved understanding of the concepts of maximum sustainable biological and economic yields, which are leading to more sophisticated management concepts.

Several decades ago many people thought fishery resources of the oceans were inexhaustible. In recent years, however, the understanding of the potential catches has been considerably modified. Research will need to continue. Three principal needs must be satisfied: to improve our understanding of the distribution and abundance of all marine life in time and space; to elucidate the environmental requirements of the resource; and to eliminate or alleviate the social–political impediments to efficient harvesting.

The methodology for managing stocks of fish, developed in the science of population dynamics of harvested fish populations, has made its greatest advances in the last 20 years. These advances are incorporated in three basic ideas: the yield–per–recruit idea; the equilibrium–yield idea; and the stock–recruitment idea.

The yield–per–recruit approach to management can be thought of in terms of all the fish in a population born in any particular year. At the time an individual fish becomes susceptible to capture, it is called a recruit. The initial number of fish in this group will continually diminish by natural mortality from a variety of causes. During the early life of this group of fish the growth of individuals will be rapid and the total weight of the group will continue to increase even though the numbers of fish are continually decreasing. As time progresses, however, growth will slow down and eventually the total weight of the group will diminish. It is clear that the total weight of the group increases, then decreases, and that the maximum yield from the population could be taken when the population is at its maximum weight. It is not, however, practical to catch all the fish at a single instant of time, and the yield–per–recruit theory tells us how to budget fishing intensity over the life-span of the fish to maximize the yield–per–individual after it enters the fishery.

The equilibrium–yield idea acknowledges that the capacity of a population of fish to reproduce itself depends on its size, and that the maximum reproduction in weight occurs at some level of population size that is intermediate between zero population size and some maximum upper limit of population size. When this maximum reproduction in weight occurs the biomass will tend to increase, but, if the "surplus production" is harvested, the population will remain in equilibrium and the surplus production can be harvested indefinitely.

The third idea, that of stock and recruitment, says that the number of individuals recruited into a fishery will be a function of the size of the spawning stock. Thus, when the spawning stock is zero, the recruitment will obviously be zero. As the spawning stock increases the recruitment from the spawning stock also increases. In some instances it appears that for some levels of spawning stock the recruitment is at a maximum, which is a desirable management criterion.

The three ideas can be used to give advice on the "optimal" amount of fishing intensity to apply to a stock. In the first the intensity of fishing is regulated to maximize the yield–per–recruit, in the second it is regulated to keep the population at a level at which the maximum equilibrium yield can be taken, and in the third it is regulated to maintain a reproductive stock which will maximize recruitment.

The importance of these population dynamics theories to human welfare lies in the fact that, while the fishery resources of the ocean are large, they are also exhaustible and it is therefore necessary to manage them. The present catches of fish amount to about 60 million tons/year. It might be that proper management of these resources, through the application of the principles of population dynamics, would yield this catch with perhaps 25 % reduction in fishing effort, thus allowing a reallocation of effort to new stocks, and thus increase man's use of the sea.

The constraints against the full application of population dynamics theory to the management of fisheries are severalfold. Better population dynamics models are needed. Exchange of research results and discussion of their application are needed by all countries fishing international waters. The knowledge provided by fishery science must be applied by international agreement.

DISEASES AND PARASITES

Man is not alone in his consumption of aquatic animals. As with other forms of life, fish and shellfish are susceptible to predation and disease. Research concerned with diseases, parasites, and genetics of aquatic animals is not extensive. Much of the research in these fields has resulted from medical and public health needs. Yet significant contributions to human welfare can be identified.

In the United States and elsewhere in the world there are extensive systems of public and private trout and salmon hatcheries. Prophylaxis and treatment of diseases has been an important part of the technology developed for such systems.

The developing knowledge about certain virus and protozoan diseases of fish in Europe and Asia has led to steps preventing the importation and dissemination of the pathogens in the United States, and has helped to exclude diseases which could devastate hatchery and wild populations of salmon and trout. Knowledge developed during the past decade about the ecological requirements of protozoan and fungal pathogens of oysters has enabled the shellfish industry to modify its operations, and thus prosper even in the presence of the pathogens.

Certain parasites of aquatic animals have received attention because they are potentially infective to humans. Outstanding are the schistosome worms that invade the blood stream. They are transmitted by freshwater snails. Schistosomiasis research has resulted in eradication of the snail vectors from many areas and in the development of prophylactic ointments and drugs to treat human cases. These measures, plus an extensive education program, have resulted in drastic reductions in prevalence and impact of the disease on human populations.

A number of intestinal flukes and tapeworms that parasitize man use aquatic animals as intermediate hosts. Research has identified intermediate hosts and the geographic foci of infection, has developed measures to prevent reinfection of fish hosts, and has led to educational programs to prevent human infection by proper processing of fish.

Certain larval nematode parasites of fish may cause human disease. Although the extent of the problem has only been recognized in the past several years, research has already resulted in recommendations for processing of fish that would completely eliminate the danger of human infection.

One of the outstanding examples of how genetics

research with aquatic animals can contribute to human

welfare is the remarkable improvement of a race of Pacific salmon by severe artificial selection. A race of fast-growing, large-sized, and rapidly maturing fish has been developed at the University of Washington, and this work serves as a model for similar artificial modifications in other species of aquatic animals.

The matter of disease resistance in aquatic animals has been studied in both fish and invertebrates, and some artificial selection for resistance to particular pathogens has been done. With oysters, evidence has been acquired that early exposure of seed oysters to pathogens may confer a form of acquired resistance. This information is important to the management practices and operations of the shellfish industry.

MAN AND HIS ENVIRONMENT

Research on pollution problems in the marine and brackish water environment is directed toward the amelioration of man's existence. The importance or immediacy of some research programs is sometimes obscured by their indirect approach to problems, but most marine pollution research is associated with the preservation of significant sectors of *1*) man's environment, *2*) his health, and *3*) his food supply.

Gross pollution of the marine environment is caused mostly by the intentional use of the sea as a waste receptacle. The vastness of the ocean has given a false sense of its capacity for assimilating wastes. The obvious degradation of water quality in some estuaries indicates that changes could eventually take place in coastal waters further from shore. In addition to deliberate dumping, enormous amounts of waste material are contributed by run-off from drainage basins, and wastes received through atmospheric fallout constitute an unknown but probably important contribution.

Standard methods have been established for the evaluation of pollution levels at which plants and animals can live. This has resulted in the definition of criteria for water quality and determination of minimum standards of acceptability.

Bioassay techniques have been developed to monitor environmental changes. Analyses of shellfish reveal the presence of persistent chemicals and heavy metals in the surrounding water. Development of automated sensing devices make possible the continuous monitoring of factors important to the health of the marine environment.

Less clearly identified with the investigation of pollution problems, but nonetheless essential, are programs which establish the norms of existing aquatic ecosystems. Such studies identify changes in the numbers and kinds of animals and plants that may occur in succeeding years as a result of pollution.

Medical research has demonstrated the relationships between contaminated seafood and human disease. The role of shellfish as carriers of organisms causing viral and bacterial diseases is well known. Oysters and clams concentrate pathogenic microorganisms from their marine habitat. Consumption of infectious shellfish accounts for about 200 cases of infectious hepatitis in the United States each year. This is not a large number from the standpoint of public health, but it causes loss of revenue from condemned shellfish beds. Shellfish concentrate infectious agents such as polio and other disease-producing viruses as much as 1,000 times over the concentration found in surrounding waters. Some of these viruses may survive in salt water for as long as 2 months, and it has recently been demonstrated that cooking methods such as frying, boiling, and stewing do not render some viruses in oysters noninfectious.

Much effort has gone into studies of shellfish biology and the water quality requirements essential for the production of clams and oysters of acceptable quality. The seeming inevitability of further pollution of shellfish-growing areas has also prompted research on depuration, the process in which contaminated shellfish cleanse themselves when they are maintained for a few days in flowing sterilized seawater. Another approach to protecting man from contaminated seafood is to insure that such products are not harvested. Extensive monitoring is conducted of fish and shellfish, as well as of estuarine water and sediments, for the presence of harmful residues of persistent organic chemicals, heavy metals, and bacterial pollution. The dependence of a majority of commercial fisheries on the estuarine zone gives special importance to pollution abatement in this area.

The success of pesticides in assisting man in his efforts to produce food on land is offset by their hazard to non-target animals, including aquatic food resources. Some of the more significant research contributions in the field of pollution biology are the studies on the kinetics of persistent synthetic chemicals in aquatic food chains, and of the accumulation of residues that are toxic to important species in man's food supply.

To ensure the continuation of man's harvest from the sea and the maintenance of an esthetically pleasing environment, research in the field of pollution biology must be greatly expanded. Man must abandon the concept of the ocean as a limitless repository for all wastes. Research must be intensified to develop methods for recycling wastes and to insure that only innocuous residues are returned to the environment.

HEALTH AND MEDICAL RESEARCH

Prior to the last 25 years, a few scholars worked on marine microbiological problems of individual interest, but during and after World War II concepts and methodology advanced, integrating the work of marine microbiologists into the whole of marine science. This new activity has permitted marine microbiology to make significant contributions through studies on *1*) properties of microorganisms that enhance our understanding of the origin and processes of life, *2*) microorganisms as they relate to properties of water masses, *3*) effects of microorganisms on productivity of the sea, *4*) stimulatory or inhibitory effects of microorganisms on other forms of life, and *5*) transmission or causation of disease to other forms of life by waterborne microorganisms.

The preponderance of evidence about the origin of life suggests that some process occurred in the primeval seas where nucleic acids or genetic materials were first synthesized. The simplest form of nucleic acids in nature occurs in the virus, and these nucleic acids are surrounded by a protein coat. However, there is evidence that strands of nucleic acid without protein coats may also exist. Further genetic investigations of microbial forms of the sea may fill the many gaps in our knowledge of primitive genetic material organization. Less than 10 viruses of marine animals have thus far been isolated, compared to more than 200 viruses of terrestrial life forms.

Many marine algae and invertebrate animals produce chemical substances that are essential to their growth and development, but are toxic or injurious to other forms of life. Man has exploited the microorganisms of the soil and produced a vast array of lifesaving antibiotics; he has only begun his search for therapeutic drugs from the sea, but already several substances of medical importance have been identified and characterized.

Carrageenan, a sulfur-containing polysaccharide extracted from seaweed, has proved useful in ulcer therapy. Oral doses afford protection from and aids in healing both gastric and duodenal ulcers. Carrageenan also acts as an anticoagulant, and seaweed hydrocolloids have been used as antilipemic agents. Kainic acid extracted from a red seaweed has been used successfully in Japan against a parasitic round worm, a whip worm, and the tape worm.

Extracts from a variety of sponges show broad-spectrum antibiotic effects, especially against human pathogens such as staphylococci, pseudomonas, acid-fast bacteria, and certain pathogenic yeasts. The exact chemical nature of most of these compounds has not yet been determined.

The D-arabinosyl nucleotides, spongothymidine, and spongouridine, were isolated from the West Indian sponge, *Cryptotethya crypta*. These compounds served as models for the synthesis of D-arabinosyl cytosine, a synthetic antiviral agent. "Paolins" extracted from abalone, oyster, clam, queen conch, and sea snail have been reported to be both antibacterial and antiviral. Paolin I inhibits the growth of several viruses in tissue culture and has protected mice infected with poliomyelitis, influenza, or polyoma virus.

Eledoisin, a polypeptide obtained from the posterior salivary glands of the octopus, is 50 times more potent than acetylcholine, histamine, or bradykinin in producing experimental hypotension. Its usefulness in treating high blood pressure in man is currently under investigation. Holothurin, a steroid saponin extracted from the Bahamian sea cucumber, suppressed the growth of certain tumors in some strains of mice.

Vast numbers of algae and other marine organisms contain substances that may contribute to control of human diseases, but the majority of these have yet to be studied in detail.

Living organisms of all kinds maintain a unique internal ionic composition that differs characteristically from the composition of the environment. In most animals this is reflected in higher intracellular concentrations of potassium and lower concentrations of sodium. Many functional activities such as muscle contraction, transmission of nerve impulses, secretion by some glands, activity of the kidney tubules, and absorption of digested food products depend on this concentration difference across cell membranes. The study of these basic cellular phenomena has been particularly advanced by the use of marine organisms.

The concept of sodium exclusion (sodium is kept out of the cell by a differentially permeable cell membrane) was first established in studies of the marine alga *Valonia*. Subsequent studies of the enormous nerve cells located in the stellate ganglion of the squid and some of its relatives showed that some sodium exclusion also characterized cells of the nervous system. Studies conducted on the giant nerve fibers of squid have provided almost all that is known about the mechanisms of conduction in the nerve fiber. The axons in the squid mantle nerves are nearly a millimeter in diameter, with a cross-sectional area 100 to 1,000 times larger than that of a single human nerve fiber. The size alone of these giant axons permits manipulations that would be impossible in other nerve fibers. Experiments have shown that the transmission of the nerve impulse results from a wave of changing permeability to sodium and potassium that sweeps over the nerve fiber, causing changes in electrical potential.

No single animal has been more important to the development of our understanding of renal function than the angler fish *Lophius piscatorius*, whose small body contains a kidney that lacks glomeruli (the filtering apparatus). One of the major contributions to renal physiology was the demonstration that urine can be produced without an initial filtration process. The discovery of the unique attributes of the angler fish focused attention on the requirements for determining glomerular filtration rates and thus directly contributed to the development of the methods now most widely used by human clinicians to determine kidney function.

Studies have clearly described the mechanics and thermodynamics of muscle contraction, using marine and terrestrial animals. The giant barnacle, *Balanus nubilis*, contains single muscle fibers up to 10 cm long. It has been shown that in this giant fiber the electrical potential of the contracting muscle is due only to calcium flux. This observation on this marine organism has sparked a series of researches in higher animals, establishing the importance of the calcium ion in the contraction of both skeletal and cardiac muscle.

The visceral organs of many fishes of the puffer group contain a virulent poison, tetrodotoxin, that is responsible for much human distress and some fatalities each year. Pharmacologists have identified the mode of action of this material, and chemists have determined its molecular structure. Tetrodotoxin owes most of its effectiveness to its ability to inhibit the ionic active transport processes on which nerve conduction, muscle contraction, and kidney tubule function depend.

The ready availability in nearly limitless numbers of ova and spermatozoa of marine animals has made possible tremendous progress in the understanding of the biochemistry of development. Gametes from marine organisms have provided raw materials for the extraction, purification, and characterization of DNA, the fundamental substance of heredity. The concept that the amount of DNA in the nucleus is directly proportional to the number of chromosomes was developed in a marine organism and provided the start for many studies of the role of the nucleus in cellular synthesis and growth and thus provided useful tools for the development of modern biochemical genetics.

The gametes of marine organisms have also provided raw materials for the definitive studies of the mechanics of the division process, by which cells multiply and by which cellular ingredients are partitioned among daughter cells. These processes are fundamental to an understanding of the orderly development of higher animals.

Fishes supply excellent clinical subjects for the study of activities helpful to man's well being. There are at least 30,000 species of fishes, as compared to 5,000 species of mammals. Fishes live under a wide range of ecological conditions such as temperature, salinity, pressure, and acidity. Like many mammals, they make excellent animals for biomedical research. Fish show behavior patterns of high order. Their body plan, tissue architecture, and physiological functions are essentially the same as in mammals. Some fish are vegetarians, while others are meat eaters. They all require proteins, carbohydrates, and fats for nourishment, and they need essential amino acids, vitamins, hormones, and minerals for proper growth. They are also subject to stresses not unlike some of those that affect man. They are susceptible to diseases, many of which are counterparts of human ailments, such as viral and rickettsial diseases, tuberculosis and various fungus diseases, trypanosomiasis and schistosomiasis, tumors and other cancers, and to environmental stress and pollution. In addition, fishes develop vitamin-deficiency diseases such as polyneuritis and anemia, degenerative diseases such as cirrhosis and fatty degeneration of the liver, and many metabolic disorders, including cataracts, gall and urinary stones, and diabetes. Even the inflammatory responses are similar in many respects to those seen in mammals. Fishes also get old, and if not eaten by predators, die from natural causes.

Fishes are good assay and test animals for pollution and for such biologics as toxin, hormones, vitamins, antimetabolites, and antibiotics. The aggressiveness of Siamese fighting fish and the pigmentary complex of most fish are built-in systems for analyzing cholinergic and adrenergic drugs. There are hundreds of publications on fishes dealing with evolution and genetics, physiology and pharmacology, histochemistry and cytochemistry, parasitology and pathology, reproductive and behavioral studies, and even in aerodynamics!

Of even more direct medical importance is the study of the mechanisms by which organisms are made aware of changes in their environment. Receptor physiology has been investigated in a variety of marine animals, and the results of these studies have been applicable, almost without modification, to man. For example, investigations of the biophysics and biochemistry of vision in the horseshoe crab *Limulus* have done much to clarify the fundamental problem of photoreception and the relationship between incident energy and the generation of the nerve impulse in the optic nerve. These results have direct application to the visual processes in man.

Such basic studies conducted on marine organisms have all contributed significantly to our understanding of normal human function. They add to the long list of contributions of aquatic biology to the welfare of man.

SELECTED ADDITIONAL READING

1. CHAPMAN, W. McL. Seafood supply and world famine—positive approach. AAS Symp. *Food from the Sea.* 29 Dec. 1969.
2. FIRTH, F. E. (editor). *The Encyclopedia of Marine Resources.* New York: Van Nostrand Reinhold, 1969.
3. GILBERT, DE W. (editor). *The Future of the Fishing Industry of the United States: A Symposium.* Seattle: Univ. of Washington, Publication in Fisheries NS 4, 1968.
4. HARDY, A. D. *The Open Sea.* London: Collins, 1959.
5. IDYLL, C. P. *The Sea Against Hunger.* New York: Crowell, 1970.
6. RICKER, W. E. Food from the sea. In: *Resources and Man.* San Francisco: Freeman, 1969, p. 87–108.
7. STANSBY, M. E. (editor). *Industrial Fishery Technology.* New York: Reinhold, 1963.

Chapter IX

Natural resources

FREDERICK SARGENT II, M.D.

BIOLOGICAL SCIENCES are concerned with all organisms including microorganisms, plants, animals, and man. The dramatic contributions of biology and medicine to man's welfare have obscured the often forgotten fact that man depends on the natural resources of the earth for this survival. Natural resources are the materials and capacities of the environment that are useful to man. Broadly defined, these resources include land and water, plant and animal life, mineral reserves, aesthetic qualities, and other geophysical characteristics of the earth. Resources may be further classified as renewable and nonrenewable. The former can be replaced or replenished by careful use and management; for example, water, timber, and range. The latter are diminished by use; for example, fossil fuels such as coal and oil.

MAN AND HIS RESOURCES

Man has long sought to dominate and control the vast resources of the earth. Every civilization has been built on interrelationships with nature involving exploitation of the renewable and nonrenewable resources at man's disposal. Use and abuse of natural resources have characterized the development of human society since prehistoric eras. Empires and nations prospered and declined in proportion to their collective successes in channeling resources to human satisfactions. Evidence is accumulating that the decay of certain past societies was foreshadowed or accompanied by environmental deterioration.

The current awareness of environmental quality has its roots in this study of man's dependency on his natural resources. Only recently has man come to realize that his welfare is inexorably linked to the conservation and effective management of air, water, land, and vegetation. As cultural evolution moved man to widespread urbanization, his dependence appeared to lessen, but in actuality this independence is illusionary.

It is difficult to separate the benefits of biological research that specifically relate to natural resources alone. Land, water, plant, and animal life, indeed most renewable resources, provide man with food and fiber. Nonrenewable resources have been utilized by man to enhance his quality of living by providing shelter, fuel for his activities, and tools and structures. Indirectly all human activity has relation to use or exploitation of natural resources. The triumphs of modern medicine take place in laboratories and hospitals kept comfortable by burning of fossil fuels. Indeed, most of the biomedical research of the past 30 years has been accomplished within buildings—newly constructed or refurbished—made from iron ore, sand, lime, water, and the products of photosynthesis. Rare earths are used in high quality instruments and equipment that are indispensable to scientific research and technology. Even the plastics industry is dependent on natural resources for raw materials or for fuel that supplies energy for production.

LAND: THE GREATEST RESOURCE

Man's greatest resource is the place where he lives, the continents. Land, with its timber forests, cultivated fields, ranges, mineral resources, and waters has been the foundation of man's activities. Classification of natural resources in terms of land types and the ecosystems that each supports is useful in attaining some perspective concerning the resources on which man ultimately depends. Most people think of land as urban or rural. Rural land is thought of in terms of pastoral dairy farms or expansive wheat fields. Many school children but few adults recall that little more than one-third of the earth is land (34 billion acres, or 38% of the surface area). Of this land mass, 10% is used to produce cultivated crops and 28% is forested. Most biological research, other than that focused on man directly, is almost exclusively concerned with the cropland and forests. Fifteen percent of the land mass is covered with polar ice caps, permanent snow, and fresh water and has not been studied as intensively as the other types of land. Few people realize that the remaining 47%—approximately one-half of the land surface—is primarily rangeland. Excluding the truly barren deserts, rangelands include areas that are too steep, dry, shallow, sandy, wet, cold, hot, or saline for cultivated crops or natural stands of timber. Its vegetation, from arctic tundra to tropic savanna and desert scrub, is the source of most of the world's meat, milk, hides, wool, and other animal by-products.

Agricultural lands. Man's closest association with land, beyond that on which he lives, is his association with agricultural lands. This land is a basic resource because on it man can produce the vast array of plant and animal food crops that he requires to sustain life.

Since the early 1900's researchers in the land grant institutions and the Soil Conservation Service of the United States Department of Agriculture have classified privately owned agricultural lands according to uses that can be sustained through generations without loss of innate productive capacity. With this knowledge of soil type, groundwater supplies, and other characteristics, millions of acres of land have been reclaimed and used for crop production.

Analysis of soil characteristics also has revealed that about 44% of the land in the 48 contiguous states is unsuited for ordinary cultivation and about three-fourths of this (642 million acres) is permanent grazing land, largely range. Its vast area offsets its low per acre productivity. Its significance to urbanites is not generally recognized beyond its value as recreational land. But more important is the fact that these areas can be used as range where vegetation is converted to protein by grazing animals.

The great expansion of automated feedlot operations in this country has been interpreted by a few as reducing the importance of these low producing nonarable rangelands. This view overlooks the important fact that livestock in feedlots originate from breeding herds that forage on rangelands. In the southern states these breeder herds forage year round. In the northern states, cattle forage 6–12 months and are fed regrowth or native hay, crop stubble, or hay cut from rangeland during winter months. Moreover, if prices of grains advance in response to demands for direct human consumption as population increases, the demands on rangelands for forage can be expected to increase. Often overlooked is the fact that ruminants in particular have the ability to convert into protein for human consumption the forages produced on lands unsuitable for the cultivated crops. In many rural areas, the suitability of the agricultural economy and man's survival are tied to continued use of rangeland for domestic animals.

Rangelands. The contributions of range science in supplying man's needs cannot be expressed in terms of national production data because most records do not enumerate production by grazing animals. Data on rangeland acreages are not provided by Federal or State crop reporting services. In addition, these census tallies of acreage do not differentiate between acres of native range and acres of tame pastureland. The latter are generally seeded with domesticated or introduced forage plants, fertilized, and periodically reseeded or renovated in various ways.

In Australia a government agency launched far-reaching research specifically on rangeland problems in the late 1960's. Near Johannesburg in South Africa research has been conducted in the ecology, agronomy, physiology, and biology of grassveld since 1931. A 1969 summary states that, of the many contributions to knowledge made by researchers at this research station, the most important to the health and well being of man are undoubtedly those connected with the more efficient use of natural grassland in providing high protein foods in the form of meat and milk.

Rangeland flora. Range ecologists were among the first
to recognize a need for complete ecosystem study and
analysis. Traditionally, they were concerned with both
plant ecology and animal ecology on rangelands. Range
studies and range management curricula were first put
on an ecosystem basis in the late 1960's. Systems analysis
has been used in the first two biome studies funded by
the International Biological Program—the Grassland
Biome and the Desert Biome. Many long-term rangeland grazing studies established 20–50 years ago lend
themselves to ecosystem analysis, although systems analysis was not one of the objectives of the original plan.

Early work on the use of browse (forage from woody
species) included taxonomy, autecology, and the development of broad guides to sustainable utilization by
grazing animals. Early work in this country, in Australia,
and in Africa described useful fodder trees. This was
followed by synecological studies of desert-shrub ranges,
and the seasonal nutritive content of shrubs. Shrub
physiology and morphology have been more carefully
studied recently.

Range scientists have investigated the control of woody
weeds for several decades. Before World War II these
control studies were directed to the use of mechanical
control methods and contact herbicides such as oil, kero-
sene, and arsenates. In recent years organic herbicides
have been developed. These herbicides have been used
in weed control and intensive agriculture, in manipulating plant succession for forestry, hormone control of
ornamental and crop plants, and defoliation for machine
harvesting of crops and for other purposes.

Rangeland fauna. Prior to 1945, many zoologists had
done descriptive ecological work on the life history, food
habits, and habitats of small mammals (mice, ground
squirrels, prairie dogs, pocket gophers, hares, and rabbits) that lived on rangelands. The common belief was
that these small mammals competed with domestic animals for forage and contributed to range deterioration.
Research suggested that their greater abundance on
depleted ranges was the "result" rather than the cause
of deterioration, while other studies pointed to the fact

that range restoration was inhibited by their foraging and burrowing.

Since 1945, many quantitative studies of small mammals have shown more fully their role on rangelands. The microscopic technique for determining the dietary habits of herbivores has shown a degree of dietary overlap between small mammals and domestic livestock in certain situations. Most mice and some squirrels are definitely of value to livestock production on grasslands since they eat insects and other arthropods that compete with livestock for forage. Small herbivorous mammals have been found to influence the occurrence, distribution, and abundance of plants on rangelands. These influences are not always in opposition to proper management of rangelands. For example, on some rangelands, moderately deteriorated by overgrazing by domestic livestock, rabbits and hares are known to exert forces favoring desirable plant species.

The new ecosystem approach to rangeland management recognizes the balancing forces between the components of the natural rangeland ecosystem. High-speed computers and new statistical schemes make possible determination of the influence of small mammals on rangelands. Population densities of small mammals on rangelands are a function of factors such as kind, abundance, and diversity of food; the occurrence and distribution of proper habitat; and the amount of predation, parasitism, and disease operating on the populations. By manipulating one or some of these factors, longer lasting biological control of populations can be achieved. Studies generally show that when rangeland vegetation is managed on a sound ecological basis the populations of small mammals and their damage to rangelands are economically insignificant, and the mammals can be used to monitor several aspects of environmental quality.

Since 1945 there has been frequent localized application of good land management practices in North America, and considerable exporting of them to underdeveloped countries. There has also been further elaboration and precision added to principles and practices relating to the hydrologic cycle and to the factors affecting the water available for beneficial use in a variety of environments. Artificial nucleation, desalinization, use of evapotranspiration suppressants, conversions of woody plant communities to water-conserving herbaceous cover, chemical treatments of watersheds for increased runoff, and land-use practices designed to achieve better use of land and water resources have received considerable attention. Watershed studies of conversions of savanna and shrubland ranges to herbaceous cover and of hydrologic effects of grazing animals are underway on rangelands. Hydrologic effects of conversions of woody riparian vegetation to herbaceous cover are being studied. As a result, the nature of the physical forces in rainstorms, grazing animal hoof-print impacts, and kinds and amounts of plant cover as they relate to infiltration, runoff, and soil movement are better known. Since the early 1960's, esthetics, wildlife values, and pollution aspects of range- and cropland treatments are receiving more attention.

FOREST AND TIMBER RESOURCES

During the past several decades advances in forest science have paralleled those associated with crop- and rangelands. Notable advances have come from application of basic research on timber production and forest management practices.

Fire was once thought to be the most destructive event in the forest. In some areas, foresters now use fire as a management tool, particularly in southern pine plantations. Such timber farms, either planted or natural, are successional to hardwood stands in the absence of fire. Thus, controlled burning under critical climatic conditions eliminates competing and undesirable hardwood species which have thinner bark, and the surviving pines grow faster.

Tree farming for pulp and dimension stock has been greatly influenced by the application of operations research techniques. In addition to the economics of the market place and new harvesting technologies, tree growth forecasts are critical input values for computer-assisted forest management. Research in forest genetics, physiology, and ecology have contributed to the ability to classify forest lands, select superior trees, and optimize woodland management.

Whole tree harvesting in order to use more of the biomass has led to the cropping of 2-year-old sprouts of rapidly growing species such as sycamore and cotton-wood. Stems up to 10 or 12 feet in height can be cut with an ensilage chopper and made into acceptable paper or particle board including stem, leaves, and bark, without any human handling. This concept has been called "silage sycamore" and permits better use of rich flood plain soils unsuited for the traditional pulp and paper tree, pine.

Forest pests. Knowledge of the biology of individual forest tree species along with the biology of the fungi have resulted in improved forest disease control measures. An outstanding example is the history of the occurrence of chestnut blight and the failure to control the effects of this disease before it destroyed the American chestnut species as a timber tree. Another disease, the white pine blister rust, was not so drastic in its abilities to spread and did not affect the entire white pine group. Relatively recently, investigations on the biology of this rust (studies of host-free species, host–parasite relationships, fungal parasite, and microparasitic relationships) are showing promise for checking the deleterious effects of this disease in our white pine forests. Biologically sound disease-control measures, such as breeding for increased natural disease-resistance in natural populations of western white pine, are developing realities. There are good indications that such biological research programs will yield a genetically controlled disease-resistant western white pine species and will maintain this rare tree species in the ecological niche that no other species of forest tree can replace.

Similar success is seen in tree genetics research programs developing pine strains resistant to southern fusiform rust disease. In another area, research on the biology of host–parasite relationships of *Fomes annonus* root and

butt rot diseases of northern hemisphere conifer species
is resulting in a broad spectrum of methods for limiting
the effects of these diseases in managed conifer forests.

Forest insects are yet another competition for timber
and wood products. Insects are among the most abundant forms of life in the forest. These organisms exert
a continuous influence through all stages of the growing
and mature forest, literally from the birth to the death
of the forest. For instance, some insects attack the fruit
or seed, often limiting reproduction of certain species;
some injure or kill seedlings and young trees, thus exerting a strong influence on species composition; others
frequently attack the mature forest and set the stage for
development of a new stand.

The manner in which entomological agents operate in
any forest is extremely complex. Populations of insects
are affected by a complicated web of interrelationships
with other animals and plants. To understand the population dynamics of any forest insect it is necessary to unravel such relationships as the influence of weather on
hosts and the several stages of insects as they mature,
the influence of site on host quality and quantity, and
the influence of these latter factors in insect populations.
These relationships, together with the host selection behavior of insect populations, determine the degree of
damage a given species of forest insects may cause.

These basic understandings in forest biology must be
coordinated before realistic management of insect populations is possible. Studies on the basic biology of insects
and their hosts identify the weak links where man can
intervene and effect control. Recently the biology and
host-selection behavior of several bark beetles species—
insects that destroy vast stands of conifers—have been
studied, and beetle-produced pheromones have been
shown to be a major stimulus in guiding bark beetles
to their hosts. Pheromones are chemical substances released into the environment by organisms. These recently
discovered "chemical messenger" substances influence
the behavior of other organisms of the same species
(for example, sex attractants). Several of these pheromones have been isolated, identified chemically, and

synthesized in the laboratory. Accordingly, there are several research projects in the applicability of manipulating bark beetle and other insect pests through use of synthetic attractants.

Other studies have shown that certain forest insects respond to olfactory stimuli originating from their hosts. When these compounds are identified and their relationship to insect host selection clearly understood, it may be possible to breed trees immune to insect attack.

In these ways basic biology can integrate with principles of applied entomology into more efficient methods of forest pest control. These relationships and areas of commonality among the biological disciplines are most important today. Forest biologists are faced with increasing demands for more effective ways of preventing losses caused by pests without, at the same time, creating unnecessary hazards for fish, birds, wild and domestic animals, and man. This is a natural development resulting from a rapidly growing public awareness of the importance of the multiple uses of forests.

GAME MANAGEMENT

Game populations have been the target of human activities for centuries. Man has inadvertently eliminated many species while regulation of other populations has often defied both poachers and bounty hunters. Research is enabling us to understand the complex food webs on land, in rivers and lakes, and in the sea that sustain these wild populations. We have learned that habitat manipulation is more meaningful than predator control, and that the consequences of introduced exotic species are not always predictable. We have also learned that toxic wastes of our technology may reach man via wildlife food chains. Radioactive cesium released into the atmosphere by nuclear testing has been deposited on tundra vegetation, particularly the long-lived lichens, and then grazed by caribou. Some of these caribou are then eaten by wolves and Eskimo. The concentration of cesium increases about tenfold from vegetation to caribou to wolf or inland Eskimo in Alaska. These Eskimos carry radioactive body burdens much higher than most human populations. It is this food chain concentration that causes such great concern about seemingly minute amounts of toxicants in the environment.

TECHNOLOGY IN RESOURCE MANAGEMENT

Remote sensing exemplifies an engineering technology stimulated by military and space requirements that represents a significant capability to measure and monitor biological and environmental parameters such as temperature, reproductive cycles, and heartbeat from airborne platforms or orbiting satellites. Most of these applications to resource management will not be realized until research is better able to associate the signals perceived with the actual occurrence or process on the ground that they signify. Often new questions about biological processes are posed by such coordinated ground and airborne studies. For example, consider the number of thermometers required to delineate a thermal gradient in an estuary or on a mountain side as opposed to a thermal map generated from a single overpass of an appropriately instrumented aircraft.

From investigations with new thermal, radar, and photographic instruments, several routine applications have emerged. Forest fires can now be detected with great accuracy even in the presence of obscuring smoke. In Florida the citrus industry can detect certain types of diseased trees and remove them before the infestation spreads. Similar techniques have been demonstrated for foliar diseases in potato, corn, and wheat fields. Discovery of insect or disease infestations in mixed forest stands is more difficult because of the heterogeneous canopy. At present the remote sensing techniques are being used very effectively in translating basic resource ecology into comprehensive pictures of natural landscapes, providing a basis for intelligent land use.

The need to understand the complex interactions of natural resources has emerged from studies of crop, range, and forest lands. These disciplines have provided the data for a more theoretical basis of man's interaction with earth's resources. Ecology and ecosystem analysis have provided the theoretical bases for understanding of these natural resources on which man depends.

MICROBES—THE UNSEEN RESOURCE

While much of man's effort to understand the world around him has focused on those resources he can see, the microbial world is another sphere with which biological science has been involved. Bacteria, fungi, and viruses are most often viewed as agents of human disease or organisms that compete with man for food and fiber. However, in recent years man has come to recognize the microbial world as an indispensable part of the biotic world and a tremendous natural resource. The constructive potential of this unseen world is not fully appreciated even now.

The mass of microbial life on earth has been estimated to exceed the total mass of animal life by about 20-fold. Biological research on this thriving, ever active, diverse group of microorganisms, their activities and their products has enriched the economy and the lives of individuals in scientifically advanced countries. Consider the numerous products, as reviewed by L. E. Casidas, Jr., that are of actual or possible commercial importance: *1)* antibiotics: such as streptomycin, penicillin, and the tetracyclines; *2)* organic solvents: acetone, butanol, ethanol, and amyl alcohol; *3)* gases: carbon dioxide and hydrogen; *4)* beverages: wine, beer, and distilled; *5)* foods: cheeses, fermented milks, pickles, sauerkraut, soy sauce, yeast, bread, vinegar, mushrooms, and acidulants such as citric acid; *6)* flavoring agents: monosodium glutamate and nucleotides; *7)* organic acids: such as lactic, acetic, citric, gluconic, butyric, fumaric, and itaconic acid; *8)* glycerol; *9)* amino acids; *10)* steroids; *11)* compounds used as chemical intermediates for further chemical synthesis of economically valuable products; *12)* yeasts for bread, food, and feed; *13)* bacteria that fix nitrogen in roots of leguminous plants; *14)* insecticides from bacteria, for example, *Bacillus thuringiensis*; *15)* vitamins and other growth stimulants:

B$_{12}$, riboflavin, vitamin A and the gibberellins; *16*) various enzymes; and *17*) fats.

Microorganisms in nature. About half of these products are related either to the treatment of disease or to the production of food. Some of the microbial products (e.g., vitamin B$_{12}$) are produced in small quantities while antibiotics are made by the ton. Animal feed yeasts are intended for commercial production at million-ton-per-annum levels within 5 years. Microorganisms help regenerate oxygen, fix nitrogen in the soil, return sewage and waste to usable form, and maintain the various cycles of life that constantly replenish the materials for organic growth. Paradoxically, it is medical microbiology and sanitation that have contributed most to over-population, yet food microbiology combined with the "Green Revolution" carries the promise that "excess" people can be fed. The proper control and utilization of viruses, bacteria, actinomycetes, algae, fungi, and protozoa are of greatest practical importance. Research is directed toward eliminating the microbes that cause disease and cultivating those that can be put to work for mankind. Because useful microorganisms can be bred or selected for superior productivity, it follows that the strains specifically obtained to do useful tasks are artifacts of industrial technology. They are unlike their unselected ancestors and may have been isolated through research extending back for 25 years or more. During selection, productivity may have been increased many hundreds of times. Obviously, as noted in previous chapters, the drug- and food-producing microorganisms are among our most valuable resources.

Microbes in the service of man. The process of drug manufacture usually consists of taking raw materials and converting them, step-by-step, into a final product. Each step consists of a chemical modification involving substitutions or alterations on a molecule. Some of the steps in the manufacture of certain drugs can be carried out by microorganisms or their enzymes with great efficiency and minimal expense. A classic system is steroid transformation.

The steroids are a group of substances that include naturally occurring cellular constituents with quadripartite ring-structured molecules. Familiar examples are cortisone, cholesterol, and testosterone. Steroid hormones are powerful mediators of metabolism. Modified steroid molecules serve as birth control pills or antiinflammatory agents; also, for the treatment of rheumatoid arthritis and specific types of malignant tumors. Since animal sources of steroids are insufficient for medical use, the pharmaceutical industry—led by research biochemists and botanists—have identified plants rich in steroid materials. Unfortunately, the plant steroids require modification (steroid transformation) to attain the desired pharmacological activity. Insertion of a hydroxyl group at position eleven on the ring system is usually essential in converting inexpensive plant steroids into hormonelike drugs. The task is formidable to the chemist. Biologically, it is easily performed by the fungus *Rhizopus* and other fungi. During manufacture, chemical and biological methods are carried out sequentially in going from substrate to drug product. The number of highly specific changes that bacteria, actinomycetes, and fungi can introduce into steroid molecules have made these drugs economically feasible and have produced entirely new and better drugs that do not occur naturally. Steroid transformation has now become a distinct technology. The roots of this technology are in basic research on the chemical activities of microorganisms. Strains of organisms able to promote desired chemical modifications are selected from nature and retained as resources for science and industry.

Microorganisms residing in the intestinal tract of man and animals release vitamins of the B group that can then be absorbed and utilized. Three vitamins produced in quantity by industrial fermentation are vitamins B_2 (riboflavin), B_{12}, and vitamin C. Biological research has shown that two fungi, *Ashbya gossypii* and *Eremothecium ashbyii*, are excellent riboflavin producers. A production of 180 tons/year is feasible. Vitamin B_{12} is obtained from microorganisms, using selected actinomycetes or bacteria. Vitamin B_{12} can be extracted from sewage sludge, where its presence indicates the past metabolic activity of various unicellular forms active in the processing of raw sewage prior to effluent discharge. Exceedingly small amounts of vitamin B_{12} are adequate for the control of pernicious anemia. Vitamin C, employed as a fortifier of fruit drinks and an antioxidant, is produced in great quantity.

Biological research has provided microorganisms that can be used in some chemical analyses far more cheaply and rapidly than chemical methods alone. When amino acids, vitamins, or antibiotics have been produced by microbes or otherwise, the laboratory must be able to certify the product concentration. The biological method for assaying an antibiotic takes advantage of the fact that a particular degree of inhibited bacterial growth indicates a particular level of drug. Standard test organisms are used. The level of growth obtained is a direct measure of the amount of a vitamin or amino acid in the tested material. The organisms utilized for bioassays are chosen only after considerable preliminary research on suitability and to this extent they are carefully maintained and may be considered, in some instances, as uniquely qualified.

Great effort is now being made to find antibiotics active against human cancer. Thousands of fermentation broths have been examined—particularly in the United States and Japan—with the resulting discovery of more than 100 substances showing some activity. Currently, only a few are manufactured. Some types of human cancers are suspected of being virus caused; the search for anticancer and antivirus antibiotics has common ground. A fundamental difficulty is the close biochemical relationship between normal tissue and cancer derived from it. Antibiotics active against cancer tend to be toxic to the related normal cells; the toxicity problem is characteristic of cancer chemotherapy in general.

Recent biochemical and biological research is beginning to find real enzymatic or virologic distinctiveness in cancer cells. Armed with the latest findings, there is a continuing search by microbiologists for antibiotics or microbial products that might diminish the toll of intractable cancer that is one of the leading causes of death. It is ironic that antibiotics have contributed to the high incidence of cancer, simply by extending the average life expectancy. It is hoped that the search for microbial products active against cancer cells will be fruitful, but the last chapter of this story is still to be written.

Microorganisms play a foremost role in the food chain of man, quite outside of their direct potential as protein. In association with legumes and as free living forms they are instrumental in fixing atmospheric nitrogen and increasing soil fertility. They help cattle and other ruminants to digest cellulose, thereby converting carbohydrate sources we cannot use into meat and milk. They degrade dead organic matter, returning elements to the soil. In the numerous biological cycles of nature, the microbes stand as prime agents in making the wheels turn. Even though microscopic forms of life are ubiquitous, abundant, and edible, it is unfortunate that cereals and cereal products constitute the major source of protein in the world's food supply. Traditional farming affords the easiest method to produce nourishment.

Protein from microorganisms: a new vista. Cereal proteins, in comparison with those of meat, tend to be deficient in composition; certain amino acids—notably lysine, tryptophan, and threonine—are often present at inadequate levels in rice, wheat, oats, maize and barley. As a consequence, it has become important to fortify flours and cereal grains with traces of the low-level amino acids, particularly lysine. The process is becoming economically feasible, since the quantities of amino acids required are about 0.2% or less. Use of microorganisms provides two approaches to the practical problem of fortifying cereal products. One is to manufacture the necessary amino acids directly; the second is to enrich cereal proteins by other foodstuffs that overcome the nutritional imbalance. By either procedure, microorganisms play a predominant role. The Japanese have developed a mutant bacterial strain of *Corynebacterium glutamicus* that produces lysine. The genetically variant bacteria have an inherited metabolic defect that results in the production of more lysine than would be obtained from routine bacterial fermentations.

Although most animal and plant foods consist ultimately of cellular material, the term "single-cell protein" connotes food of microbial origin, predominantly made up of bacteria, fungi, or algae treated by the technology necessary to assure palatability and nutritive value. In considering single-cell protein as a resource, it is interesting to observe that such a food supply is unlike conventional agricultural products that grow on a diminishing area of usable land. Even in the United States there are large regions unsuitable for farming or grazing, but adequate for the cultivation of food yeasts or freshwater algae. Economically, it is best to cultivate yeast in the vicinity of the raw materials used for growth (e.g., molasses), but algae can be grown wherever water, sunlight, nitrogen, air, and carbon dioxide sources are available. The practicality of recycling used cellulose products such as paper and rags to produce single-cell protein is a subject of current research.

The technology of yeast production is not new; food yeast has been used in Germany since the first World War. Millions of pounds each year are consumed in the United States, mostly as animal feed additives. In the mid-1960's, the worldwide production of yeast for animal and human food was about 5,000 tons/week. Further research on palatability is needed; yeast foods extensively tested in Costa Rica, Jamaica, and the Philippines have not received unqualified acceptance. Nutritionists have learned that cultural food habits are extremely difficult to modify. Also, economic factors may enter the equation. A world that will possibly contain 6 billion people by 1980 must modify its food habits. The use of single-cell protein affords another way to avoid malnutrition and starvation.

Although many obstacles stand in the way of using single-cell protein as a common staple in the human diet, the compelling positive argument in the face of rising population pressure is efficiency. We can use the sun's energy to grow grass or corn, feed this to animals which inevitably extract only part of the energy, then sacrifice the animals to provide a few cuts of meat, throwing some of the carcass away. This takes many months. The conversion of solar energy into algal protein is far more rapid and efficient; there are no wasteful intermediates. Using the least expensive sources of carbohydrate and nitrogen, yeast and bacteria can be cultivated rapidly in great bulk. A single bacterial cell, growing unchecked for 2 days, could vastly exceed the bulk of the earth assuming a division every 20 min and an infinite food supply. This example is impractical in operation—perhaps fortunately —but yeast, bacteria, fungi, and algae may prove, with adequate research, to be the best solution to man's most critical problem: how to feed the exploding population.

Extensive investigation has been going on for the last 10 years on conversion of petroleum fractions and crude oil into single-cell protein for animal feed and human food supplements. Some reciprocal benefit may be obtained by the petroleum engineer: the yeast *Candida lipolytica*, when growing in crude petroleum, removes certain fractions of the crude petroleum that must be eliminated in the refining process. Yeast produced on an industrial scale, using petroleum and related hydrocarbon substrates, is an active and worldwide enterprise. Some experts have estimated that a billion tons of crude oil are used by mankind each year and that converting 10% into single-cell material consisting of about one-half protein would provide 50 million tons annually of high quality food supplement. The protein deficit between now and the year 2000 could be entirely overcome by this process. The cells multiply on oil–water interfaces, artificially aerated and provided with nitrogen sources and a few simple chemicals. Many countries including Japan, Czechoslovakia, Great Britain, France, and Taiwan are in the business of obtaining protein from petroleum-related substrates variously classified as high waxy crudes, gas-oil, *n*-paraffin, flue gas, and natural gas.

The difficulties imposed by toxicity, palatability, and untested consumer attitudes should not lead to discouragement, but to greater effort by nutritionists, geneticists, and microbiologists to produce an acceptable product. Biological research has provided in the form of microorganisms a rapidly growable food source able to utilize a wide variety of petroleum by-products, in addition to inexpensive hydrocarbon and carbohydrate materials generally, for conversion into edible substances. At present, this major food resource is not adequately used.

CONVERSION OF WASTE TO RESOURCE

An ancient and indirect approach is the use of sewage or organic waste as fertilizer for food-yielding plants. Putrefactive and degradative microorganisms are essential; unless the organic material is broken into small molecules, it cannot be absorbed by plant roots. In recent years a new approach to waste utilization has been developed: Why not grow bacteria, fungi, or algae on organic waste matter or on other discard substrates, and use the microorganisms *directly* as food? Could a yeast "beefsteak" be a dietary staple? Experience shows that the problems are not technical, but aesthetic. Although highly nutritious and easy to produce on inexpensive substrates, single-cell protein may not suit the human palate. The "Green Revolution," which introduced new varieties of rice and wheat into underdeveloped countries, showed clearly that even varietal difference in rice or wheat flavor may bring disfavor from potential consumers who are already at starvation levels of nutrition. The move to use yeast algae or bacteria grown in sewage as human food requires crossing an emotional barrier, but the extent of this barrier can be diminished. Experimental modification of single-cell protein may make it more palatable.

In spite of social reluctance, we are already planning to reuse some sewage. One process, already developed and now under extensive test, runs the sewage past spinning discs which are covered with microorganisms. In the final phases of purification, algae remove nutrients from the water. The algal growth is "harvested" from the discs, with a yield of about 6 g/m² each day. Conversion of human waste into food by microbial means seems visionary but it has an attractive feature. The waste itself becomes a resource.

How are the valuable resource strains of microorganisms and other cells kept to insure their safety and stability? Most can be freeze-dried or preserved in the extreme cold of liquid nitrogen. Under such conditions, chemical changes are virtually absent. University, government, and industrial scientists maintain storage banks of drug-producing or otherwise desirable stocks of microbes. Viruses and animal cells are likewise placed in liquid nitrogen, to be removed and revived on demand. There is a worldwide network of organizations concerned with meeting the need for certified strains and species. Prominent in the United States is the American Type Culture Collection, located in Rockville, Maryland. This center, supported by Government funds, is staffed by biologists who do basic research or work on the identification, preservation, and storage of microscopic forms of life. More than 12,000 different types of bacteria, fungi, animal viruses, animal cell lines, algae, protozoa, rickettsiae, and bacteriophages are kept on hand. Requested organisms, totalling almost 15,000 cultures, are mailed each year to qualified educators or investigators.

The preservation of biological resources and the continued efficient functioning of stock centers must be assured by stable fiscal arrangements. It is important to human welfare that this collection be maintained and enlarged to include other cell types that may be useful to man.

CONSERVATION OF MAN'S ABUNDANCE

Conservation of natural resources and the science of ecology have been and continue to be closely interrelated, sharing a common origin in classical natural history and a general core of methodology, experience, and knowledge. Only in their divergent emphases are distinctions recognizably valid. Basically, each is concerned with the interactions of living systems with their biotic and abiotic environments. Historically, however, and with some exaggeration to magnify distinctions, the ecologist penetrated those interactions for their own sake (i.e., "basic" research) while students of conservation and natural resources generally investigate with an end product in mind (i.e., "applied" research). While the latter direction proved to be of direct and immediate benefit to man's many-sided needs and demands, future development was limited because of the lack of a theoretical framework with its concomitant potential for organizing seemingly unrelated facts into a logical whole that would have predictive capacity. It is this theoretical framework that the field of ecology is contributing to studies in conservation and natural resources. The results have led to significant changes in man's relationships with and his appreciation for the environment.

Science of ecology. This field of biology has changed dramatically and of late as it has increasingly addressed itself to its proper province: the interfaces between living and nonliving systems. As a result, the introduction of physical and chemical technology has revolutionized what the ecologist does in the field. The renewed recognition of physiological and behavioral causes and consequences has opened a Pandora's box of laboratory studies; and most important, an increasing awareness of the generative and heuristic power of mathematical analyses and model building, and of the capacity of the computer to both stimulate and simulate, has enabled quantum jumps that are radically transforming the discipline. Out of these kinds of stimuli the concept of the ecosystem has been developed and expanded into a system that has form and function, integrity and strategy, and that interrelates in various ways with yet other ecosystems. Significantly, man is increasingly acknowledged as an irreconcilable part of such systems inasmuch as human activity interpenetrates nearly all ecosystems. Thus it is necessary that man understand each basic ecosystem unit and interrelate each into meaningful sets.

It is in this seeing of individual pieces as interrelated parts of larger and larger wholes that the "new" ecology has made its significant contribution to the fields of conservation and natural resources. But of no less significance has been the introduction of numerous ideas such as extending the concept of taxonomic diversity as a community stabilizer to the practice of agriculture and husbandry, the concept of self-regulation of predator and prey as a potent biological substitute for chemical pesticides, the concepts of population growth and regulation to maximizing sustained yields of utilitarian food populations, the concepts of biogeochemistry to recycling of natural resources, and the concept of energy flow to agricultural and marine productivity.

The inclusive and comprehensive concept of the ecosystem entails a cosmopolitan outlook that has had significant effect at the thinking level in the field of conservation. This is well reflected in the prospectus of the Conservation Section of the International Biological Program (IBP) in which there is considerable acknowledgement of the need for international coordination because of the recognition that ecosystems themselves are coordinated. There is also increased recognition of the severe limitations imposed by jurisdictional boundaries of both political and governmental sorts, since such delimitings seldom correspond with biological (ecological) reality. For example, Lake Erie is one ecosystem owing allegiance to two international and three state governments and a multitude of government agencies of smaller political entities.

The multidisciplinary approach which ecosystem analysis demands has initiated new working relationships and consortia of persons of various competencies in focusing on particular resource problems. Several universities have already undertaken comprehensive interdisciplinary attacks on major ecosystems, integrating the expertise of limnologists, hydrographers, meteorologists, aquatic and terrestrial biologists, and, significantly, lawyers, sociologists, economists, and political scientists. The objectives of such comprehensive research are the construction of models reflecting reality and enabling sound prediction and the delineation of reasonable alternates as well as guidelines for policy decisions in resource utilization and development. It is noteworthy that precisely this approach is underway in several universities whose leadership in the fields of conservation and natural resources is universally and justifiably well recognized.

Perhaps the greatest attitudinal contribution ecology has brought to the fields of conservation and natural resources is that man is as much a part of an ecosystem as any given species of weed or wildlife, that all ecology is, in this sense, human ecology. Man has been present in nearly all ecosystems since his origin, and this must be reckoned with in both research and planning. The ecologist–conservationist–natural resource person thus acquires by default, if not by design, a mantle of social responsibility of considerable magnitude, his efforts becoming increasingly directed toward ends of practical benefit for human welfare. Contributing to the intelligent management of human ecosystems thus looms as the major responsibility of students and practitioners in these fields. Such management will involve many in new sociopolitical and ethical roles for which their past training and experience will prove inadequate and thereby provide the stimulus to alter future training.

In summary, man has devoted great energy to acquiring knowledge of natural resources and technology for his use. Indeed a resource is not conceived as such until a technology has made it available or useful. We have learned, albeit incompletely and hesitatingly, that new knowledge brings its own, if unpredictable, rewards. It is undeniable that demand for resources is pegged to the population size multiplied by the levels of material expectation. Accommodating the diverse life styles of expanding world populations to finite resources with innovative technology is not as cataclysmic as is now often touted by modern Jeremiahs or as patently probable as asserted by the new technocrats. The conflicts arising from these increased demands on resources are often global and therefore not subject to geographic displacement. We can no longer expect to "move west" to uncharted territory; even space and air have acquired value. The intensity of the demands for conflicting use of resources challenges many human values. The solutions to resource allocation, abuse, or pollution will require knowledge, new socioeconomic institutions, and far greater attention to maintenance of spaceship earth. A rapidly growing living system expends great energy on production while a mature system must devote substantial energy to survive. To maintain our life support systems, the concepts and practices of efficient use of natural resources including recycling must be exchanged for those of exploitation and expansionism.

SELECTED ADDITIONAL READING

1. CASIDA, L. E. *Industrial Microbiology.* New York: Wiley, 1968, 460 p.
2. KORMODY, E. J. *Concepts of Ecology.* Englewood Cliffs, N. J.: Prentice-Hall, 1969, 209 p.
3. LEWIS, J. K. Range management viewed in the ecosystem framework. In: *The Ecosystem Concept in Natural Resource Management,* edited by G. M. Van Dyne. New York: Academic, 1969, 383 p.
4. MATELES, R. I., AND S. R. TANNENBAUM. *Single Cell Protein.* Cambridge: Mass. Inst. Tech. Press, 1968, 480 p.
5. ROUX, E. *Grass: A Story of Frankenwald,* edited by W. M. Roux. London: Oxford Univ. Press, 1969, 212 p.
6. U.S. National Committee for the Internation Biological Program. *Man's Survival in a Changing World.* United States Participation in the International Biological Program. IBP Office, Division of Biology and Agriculture, Washington, D.C.: Natl. Acad. Sci.—Natl. Res. Council, 28 p.
7. Committee on Resources and Man. *Resources and Man.* Washington, D.C.: Natl. Acad. Sci.—Natl. Res. Council. San Francisco: Freeman, 1969, 259 p.

Glossary

Adenosine triphosphate—a chemical compound occurring in all cells and used to provide energy in body reactions

Aldosterone—a hormone from the adrenal cortex which controls sodium and potassium metabolism

Algae—a group of plants which includes the seaweeds

Angiography—a technique of taking X-ray pictures of blood vessels

Antibody—a specific substance which is produced in the body in reaction to the presence of a foreign substance and which reacts with the foreign substance in an observable way

Atherosclerosis—a cardiovascular ailment in which fatty material is deposited in the inner part of the walls of the blood vessels

Biochemistry—the branch of chemistry that deals with living organisms and life processes

Biology—the study of living things; the origin, development, structure, and functions of plants and animals

Biomedical engineering—the application of the tools of engineering in the solution of medical problems and in improvement of health care

Cataract—a condition of the eye in which the lens is cloudy and will not allow light to enter the eye

Catheterization—the insertion of a tube into a blood vessel, a duct, or a hollow organ of the body for diagnostic or therapeutic purposes

Congenital—existing at birth or before birth

349

Creutzfeldt-Jakob disease—a disease of the brain and spinal cord marked by tremors, rigidity, and emotional disturbance; it is a chronic degenerative process caused by a virus

Cyclic AMP—a chemical compound which serves as a cellular intermediate and regulates cellular processes

Cyclopropane—a gas used as a general anesthetic

Dementia—a general term for mental deterioration; the presenile form begins in middle age and is caused by cerebral arteriosclerosis

Dicumarol—a drug used for the prevention of blood clotting

Diuretic—a drug that increases the secretion of urine

DNA—deoxyribonucleic acid

Ecology—the science of organisms as affected by the factors of their environment

Efficacy—effectiveness

Electrocardiography—the recording of the electrical currents produced by the heart muscle

Electroencephalography—the recording of the electrical currents produced by the brain; brain wave recording

Endocrinology—the study of the body organs which secrete the hormones they manufacture directly into the bloodstream

Enzyme—a chemical compound which is necessary for catalyzing specific chemical reactions

Epidemiology—the science which deals with the relationships of the various factors which determine the frequencies and distributions of an infectious process, a disease, or a physiological state in a human community

Erythroblastosis, fetal—a serious hemolytic disease of the newborn with high mortality that has been shown to result from destruction in the fetus of Rh-positive red blood cells by antibodies from the blood of an Rh-negative mother

Feedback systems—regulatory systems, such as thermostats and other control devices

Geriatrics—the field of medicine that treats the problems of old age and the aging

Glaucoma—a disease of the eye marked by extreme pressure within the eye causing hardness of the eye, atrophy of the retina, and blindness

Glomerulonephritis—a kidney disease in which there is inflammation of the capillary loops in the glomeruli of the kidney

Growth hormone—a hormone of the anterior pituitary gland which controls the rate of skeletal growth and the gain in body weight

Hematology—the branch of medicine that deals with the structure and function of the blood and the blood-forming organs

Hemoglobin—the oxygen-carrying red pigment of the red blood cells

Hepatitis—inflammation of the liver

Hodgkin's disease—a painless, progressive, and fatal enlargement of the lymph nodes, spleen, and general lymphoid tissues of the body

Hormone—the chemical substance secreted into the bloodstream by endocrine glands, which has a specific effect on an organ elsewhere in the body

Hydrocephalus—an abnormal increase in the amount of cerebral fluid, resulting in dilatation of the cerebral ventricles and enlargement of the head

Hypertension—high blood pressure

Hypothalamic-releasing hormones—hormones produced in the hypothalamus which control the release of hormones from the anterior pituitary gland

Immunology—the field of biology which deals with the study of immunity, the power by which the body is protected against any particular disease or poison

Immunosuppression—the blocking of immune processes within the body so that the body will not reject transplanted organs

Insulin—the hormone produced by the islets of Langerhans of the pancreas and secreted into the blood, where it regulates carbohydrate metabolism

Isotopic tracer—a radioactive element which can be introduced into the body and followed in its metabolism and distribution by means of a Geiger counter

Leukemia—a fatal disease of the blood-forming organs, characterized by a marked increase in the number of white blood cells, and enlargement of the lymphoid tissue of the spleen, lymph nodes, and bone marrow

Lung surfactant—a substance secreted by the lung tissues and which is necessary for the maintenance of the mechanical stability of the tiny alveolar sacs of the lungs

Malocclusion—malposition of the teeth so that alignment for efficient chewing is not possible

Maxillofacial prosthesis—an artificial part constructed to repair deformities of the jaw and face caused by injury, disease, or congenital abnormalities

Metabolism—the changes that take place in tissues; the sum of all physical and chemical processes by which living cells and tissues are produced and maintained

Mongolism—a form of congenital idiocy

Mycotoxin—a toxin from a bacterium or fungus

Neoplasm—any new and abnormal growth, such as a tumor

Neural transmission—the mechanism of transmission of electrical impulses along nerves

Neurology—the branch of medicine that deals with the nervous system

Neurophysiology—the function of the nervous system

Norepinephrine—a hormone produced by the adrenal medulla

Nuclear medicine—the application of nuclear reactions in medicine and the use of radioactive elements in medicine

Ophthalmology—the branch of medicine that deals with the eye and its diseases

Orthodontics—the branch of dentistry which deals with the prevention and correction of irregularities in the teeth and malocclusion

Otorhinolaryngology—the branch of medicine that deals with the ear, nose, and throat and their diseases

Oxytocin—a hormone from the posterior pituitary gland which stimulates contraction of the uterus

Paraplegia—paralysis of the legs and lower parts of the body from disease or injury to the spinal cord

Parkinson's disease—shaking palsy or paralysis agitans; there is tremor of resting muscles, a slowing of voluntary movements, strange gait, peculiar posture, and muscle weakness

Peptic ulcer—ulceration of the mucous membrane of the esophagus, stomach, or duodenum, caused by the action of stomach acid

Periodontal disease—a disease or disorder of the tissues that surround the root of a tooth

Peritoneal dialysis—a technique of removing toxins from the blood of patients with kidney failure, wherein fluid is put into the abdominal cavity through a small puncture wound in the abdominal wall. Toxins from the blood pass into the fluid and the fluid is then removed a short time later

Pharmacology—the science that deals with all aspects of drugs

Pheromones—substances produced by insect glands which exert a powerful attraction to the opposite sex

Photosynthesis—the formation of carbohydrates from carbon dioxide and water in the chlorophyll tissue of plants under the influence of light

Physiology, comparative—the study and comparison of function in animals or plants of various species

Pituitary gland—the major endocrine gland of the body, located at the base of the brain

Prostaglandin—an active principle obtained from the prostate gland and seminal vesicles; it is a pressor, vasodilator, and stimulant to the intestinal and uterine muscles

Prosthetic dentistry—the branch of dentistry which deals with dental appliances and substitutes, such as artificial dentures

Radiology—the branch of medicine which deals with the use of radiant energy in the diagnosis and treatment of disease

RNA—ribonucleic acid

Rubella—German measles

Schizophrenia—a mental disorder in which there are disturbances of feeling, thought, and relations to the outside world

Stapes—the innermost of the small bones of the middle ear; immobility of these bones results in conduction deafness

Synapse—the region of contact between processes of adjacent neurons, forming the place where a nerve impulse is transmitted from one neuron to another

Thyrocalcitonin—a hormone secreted by the thyroid gland; it lowers the calcium level of the blood

Vaccine—any material which is used for preventive inoculation

Vascular disease—any disease of the blood vessels

Vasopressin—a hormone from the posterior pituitary gland which raises the blood pressure

Warfarin—a chemical which inhibits the clotting of the blood; when consumed by rats in small quantities over several days it results in their death

Yellow fever—a virus disease transmitted by mosquitoes and marked by fever, jaundice and albuminuria.

Subject index

Compiled by Constantine J. Gillespie